Praise for *The Wild Trees*

"[Richard Preston] has written an exciting book. . . . Once again he combines the thrill of exploration with the quirkiness of those who choose it as their lives' work. . . . This book is fascinating in its keen, inquisitive account of the redwoods' biosphere." —*The New York Times*

"[Preston] turns his gaze upward, to the realm of the towering and majestic. . . . Preston was clearly moved by his experiences—the thoughtful and engaging narrative is informed by a satisfying touch of spirituality—and he brilliantly shares his ardor for the arboreal." —*Entertainment Weekly*

"[Preston] is a science writer with an uncommon gift for turning complex biology into riveting page-turners. . . . In his rich metaphorical style, Preston makes us feel the forest undergrowth tearing at the explorers' clothes, the wind swaying the 'Treeboats' they sleep in, the bees stinging their faces as they make epic ascents of behemoths." —*The Washington Post Book World*

"Anyone who has ever climbed a tree eager to experience the magic of the world seen from its heights will be grabbed by the story Preston presents." —*The Christian Science Monitor*

"Mr. Preston is clearly in love with his redwoods, even learning to 'skywalk' so that he could climb them himself. With his hands-on approach, he brings to life the mystery and majesty of these giant wonders of the botanical world." —*The Economist*

"[Preston's] complete immersion in his subject makes for a superlative work of narrative nonfiction."   —*Discover*

"Absorbing, precise . . . Preston knows how to fold the science into the seams of his narrative, and his dry humor crops up, pleasurably, at the edges of his observations."
—Cleveland *Plain Dealer*

"[Preston] is without rivals in his ability to relate complex biological systems to a lay audience. Aside from the precious groves of redwoods, the stars of this narrative are Sillett and Taylor. Their methods of exploration may be unorthodox, but their reverence for our vanishing natural world is admirable and worth emulating."   —*Chicago Sun-Times*

"A heart-thumping page turner. Think of John Krakauer's *Into Thin Air* account of the ill-fated climbers on Mount Everest—this time in the big trees of the Pacific Coast. . . . It's all captured within the stories of the climbers, their personal turmoil and their struggles in search of the tallest trees."
—*The Charlotte Observer*

"Intensely dramatic . . . The scientific story goes beyond the discovery of the tallest trees to the astonishing biology within their canopies."   —*Seattle Post-Intelligencer*

"[*The Wild Trees*] is at its heart a science book, describing in accessible terms the wonders of this newly charted ecosystem. . . . [Preston] draws sharp, often moving portraits of his cast of modern-day explorers."   —*ForbesLife*

"[Preston] applied his gift for research to learning all he can about the giants of the forest. . . . *The Wild Trees* is the intelligent sort of nonfiction readers have come to expect from him, a book that elevates, entertains, and alas, is over far too soon."   —*Natural History*

THE
WILD
TREES

*A Story of Passion
and Daring*

RICHARD
PRESTON

RANDOM HOUSE TRADE PAPERBACKS
NEW YORK

Published in the United States by Random House Trade Paperbacks, an imprint of
The Random House Publishing Group, a division of Random House, Inc., New York.

RANDOM HOUSE TRADE PAPERBACKS and colophon are trademarks of Random House, Inc.

Excerpt from *Panic in Level 4* by Richard Preston
copyright © 2008 by Richard Preston

Portions of this work were originally published in *The New Yorker.*

Originally published in hardcover in the United States by Random House, an imprint of
The Random House Publishing Group, a division of Random House, Inc., in 2007.

Grateful acknowledgment is made to the following for permission to use both published
and unpublished materials:

MICHAEL A. CAMANN: Photograph of a copepod (anthropod) collected in the Atlas Grove.
Used by permission of Michael A. Camann.

THOMAS B. DUNKLIN: Panoramic photograph of Ilúvatar and photograph of *Lobaria
oregana,* used as the basis for line drawings by Andrew Joslin. Photographs © Thomas B. Dunklin,
www.thomasbdunklin.com. Photos adapted by permission of Thomas B. Dunklin.

R. STEVE FOSTER: Drawing of Telperion of the Humboldt Tree. Used by permission of R. Steve Foster.

SAVE-THE-REDWOODS LEAGUE: Map of the redwood forest, used in the construction of the map "The
North Coast of California" by Andrew Joslin. Used by permission of Save-the-Redwoods League.

STEPHEN C. SILLETT: Scientific Notebooks, 1999 (digital reproduction of certain pages).
Data sets displaying the 3-D structure of the crown of Ilúvatar. Used by permission
of Stephen C. Sillett.

Library of Congress Cataloging-in-Publication Data

Preston, Richard.
The wild trees : a story of passion and daring / by Richard Preston ; maps and
illustrations by Andrew Joslin.
p.   cm.
ISBN 978-0-8129-7559-8
1. Coast redwood—California, Northern.   2. Coast redwood—Ecology—California, Northern.
3. Forest canopies—California, Northern.   4. Forest conservation—California, Northern.
5. Tree climbing—California, Northern—Anecdotes.   I. Title.

SD397.R3P74 2007
585'.509794—dc22   2006048646

Printed in the United States of America

www.atrandom.com

4 6 8 9 7 5

Book design by Susan Turner

To my brother Douglas Preston
Remember that tree we used to climb when we were boys?

*Those who dwell among the beauties and mysteries of the earth
are never alone or weary of life.*

RACHEL CARSON

CONTENTS

AUTHOR'S NOTE

This book is narrative nonfiction. The characters are real and the events are factual, told to the best of my understanding. Passages in which I narrate a person's thoughts and feelings and present dialogue have been built from interviews with the subjects and witnesses, and have been fact-checked. So many incredible things happen in our world that are never noticed, so many stories never get told. My goal is to reveal people and realms that nobody had ever imagined.

RICHARD PRESTON, 2007

MAPS AND ILLUSTRATIONS
BY
ANDREW JOSLIN

OF BOTANISTS AND TREES

Botanists have a tradition of never revealing the exact location of a rare plant. Contact between humans and rare plants is generally risky for the plants. Many of the giant trees I describe in this book, as well as the groves they inhabit, have only recently been discovered, and in some cases have been seen by fewer than a dozen people, including myself. To honor the tradition of botany, I won't reveal the exact locations of giant trees or groves if these locations have not been previously published. If a tree's location has been published, or if the tree is no longer alive, then I will give its location.

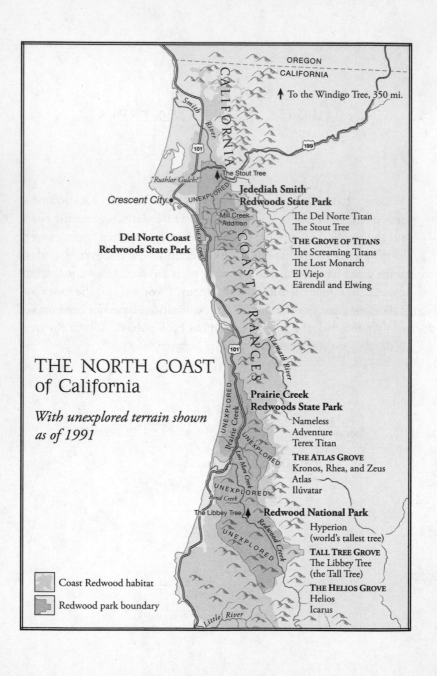

OREGON

CALIFORNIA

↑ To the Windigo Tree, 350 mi.

Smith River

101

199

CALIFORNIA

↑ The Stout Tree

"*Ruthlor Gulch?*"

Crescent City •

UNEXPLORED

**Jedediah Smith
Redwoods State Park**

Mill Creek
Addition

**Del Norte Coast
Redwoods State Park**

UNEXPLORED

The Del Norte Titan
The Stout Tree

THE GROVE OF TITANS
The Screaming Titans
The Lost Monarch
El Viejo
Eärendil and Elwing

Klamath River

C O A S T   R A N G E S

101

THE NORTH COAST
of California

*With unexplored terrain shown
as of 1991*

**Prairie Creek
Redwoods State Park**

Nameless
Adventure
Terex Titan

THE ATLAS GROVE
Kronos, Rhea, and Zeus
Atlas
Ilúvatar

UNEXPLORED

Prairie Creek

UNEXPLORED

Lost Man Creek

UNEXPLORED

Bond Creek

The Libbey Tree ↑

Redwood National Park

Hyperion
(world's tallest tree)

TALL TREE GROVE
The Libbey Tree
(the Tall Tree)

THE HELIOS GROVE
Helios
Icarus

UNEXPLORED

Redwood Creek

Coast Redwood habitat

Redwood park boundary

Little River

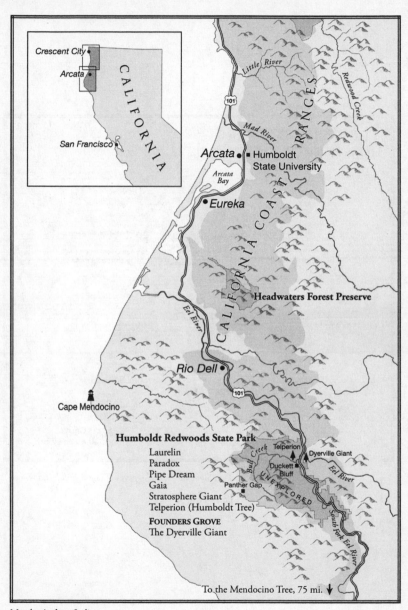

Map by Andrew Joslin

The Skeleton Forest
Hume Plateau, Victoria, Australia

Harper Creek

Position doubtful

Position doubtful

Position doubtful

GREAT DIVIDING RANGE

Dry Hills

David Ashton Tree

Michael-Taylor's Ascent

Amabilis

Big Ash One — Tallest tree
in mainland Australia

Mt. Disappointment

Running Creek

York Town Hill

The Cascades

Jack Creek

Joey Creek

Tourourrong Reservoir

To the "Bedless
Bungalow"

Kinglake West

C725

Whittlesea-Kinglake Road

Whittlesea

C725

Whittlesea-Yea Road

Contour interval = 30 meters

Hume Plateau

Melbourne

Bass Strait

TASMANIA

The "Skeleton Forest"— groves of tallest *Eucalyptus regnans*

Locked gate

Map by Andrew Joslin

1

VERTICAL

EDEN

NAMELESS

ONE DAY IN THE MIDDLE OF OCTOBER 1987, A BABY-BLUE
Honda Civic with Alaska license plates, a battered relic of
the seventies, sped along the Oregon Coast Highway, moving south on the headlands. Below the road, surf broke around sea
stacks, filling the air with haze. The car turned in to a deserted parking lot near a beach and stopped.

A solid-looking young man got out from the driver's side. He had
brown hair that was going prematurely gray, and he wore gold-
rimmed spectacles, which gave him an intellectual look. His name
was Marwood Harris, and he was a senior at Reed College, in Portland, majoring in English and history. He walked off to the side of the
parking lot and unzipped his fly. There was a splashing sound.

Meanwhile, a thin, tall young man emerged from the passenger
side of the car. He had a bony face, brown eyes, and a mop of sun-
streaked brown hair, and he wore a pair of bird-watching binoculars
around his neck. Scott Sillett was a junior at the University of Arizona, twenty-one years old, visiting Oregon during fall break. He took

up his binoculars and began to study a flock of shorebirds running along the surf.

The interior of the Honda Civic was made of blue vinyl, and the back seat was piled with camping gear that pressed up against the windows. The pile of stuff moved and a leg emerged, followed by a curse, and a third young man struggled out and stood up. "Mardiddy, this car of yours is going to be the death of us all," he said to Marwood Harris. He was Stephen C. Sillett, the younger brother of Scott Sillett. Steve Sillett was nineteen and a junior at Reed College, majoring in biology. He was shorter and more muscular than his older brother. Steve Sillett had feathery light-brown hair, which hung out from under a sky-blue bandanna that he wore tied around his head like a cap. He had flaring shoulders, and his eyes were dark brown and watchful, and were set deep in a square face. The Sillett brothers stood shoulder to shoulder, looking at the birds. Their bodies were outlined against decks of autumn rollers coming in, giving off a continual roar. Scott handed the binoculars to his younger brother, and their hands touched for an instant. The Sillett brothers' hands had the same appearance—fine and sensitive-looking, with deft movements.

Scott turned to Marwood: "Marty, I think your car should be called the Blue Vinyl Crypt. That's what it will turn into if we fall off a cliff or get swiped by a logging truck."

"Dude, you're going to get us into a crash that will be biblical in its horror," Steve said to Marwood. "You need to let Scott drive." (Steve didn't know how to drive a car.)

Marwood didn't want Scott's help with the driving. "It's a very idiosyncratic car," he explained to the Sillett brothers. In theory, he fixed his car himself. In practice, he worried about it. Lately Marwood had noticed that the engine had begun to give off a clattering sound, like a sewing machine. He had also become aware of an ominous smell coming from under the hood, something that resembled the smell of an empty iron skillet left forgotten on a hot stove. As Marwood contemplated these phenomena and pondered their significance, he wondered if his car needed an oil change. He was fairly sure that the oil had been changed about two years ago, in Alaska, around the time the license plates had expired. The car had been driven twenty thousand miles since then, unregistered, uninsured, and un-maintained, strictly off the legal and mechanical grids. "I'm worried you'll screw it up," he said to Scott.

Steve handed the binoculars to his older brother and climbed into the back of the Blue Vinyl Crypt. "Dudes, let's go," he said. "We need to see some tall redwoods."

They planned to go backpacking in one of the small California state parks that contain patches of ancient coast redwood forest. None of the young men had ever seen a redwood forest. Steve seemed keyed up.

THE COAST REDWOOD TREE IS AN EVERGREEN CONIFER AND A MEMBER of the cypress family. Its scientific name is *Sequoia sempervirens*. It is sometimes called the California redwood, but most often it is simply referred to as the redwood. No one knows exactly when or where the redwood entered the history of life on earth, though it is an ancient kind of tree, and has come down to our world as an inheritance out of deep time. A redwood has furrowed, fibrous bark, and a tall, straight trunk. It has soft, flat needles that become short and spiky near the top of the tree. The tree produces seeds but does not bear flowers. The seeds of a redwood are released from cones that are about the size of olives. The heartwood of the tree is a dark, shimmery red in color, like old claret. The wood has a lemony scent, and is extremely resistant to rot.

Redwoods grow in valleys and on mountains along the coast of California, mostly within ten miles of the sea. They reach enormous sizes in the mild, rainy climate of the northern stretches of the coast. Parts of the North Coast of California are covered with temperate rain forest. A rain forest is usually considered to be a forest that gets at least eighty inches of rain a year, and parts of the North Coast get more than that. A temperate rain forest has a cool, moist, even climate, not too hot or cold. Redwoods flourish in fog, but they don't like salt air. They tend to appear in valleys that are just out of sight of the sea. In their relationship with the sea, redwoods are like cats that long to be stroked but are shy to the touch. The natural range of the coast redwoods begins at a creek in Big Sur that flows down a mountain called Mount

Mars. From there, the redwoods run up the California coast in a broken ribbon, continuing to just inside Oregon. Fourteen miles up the Oregon coast, in the valley of the Chetco River, the redwoods stop.

The coast redwood is the tallest species of tree on earth. The tallest redwoods today are between 350 and close to 380 feet in height—thirty-five to thirty-eight stories tall. The crown of a tree is its radiant array of limbs and branches, covered with leaves. The crown of a supertall redwood has a towering, cloudy form, and the crowns of the tallest redwoods can sometimes look like the plume of exhaust from a rocket taking off.

Botanists make a distinction between the height of a tree and its overall size, which is measured by the amount of wood the tree has in its trunks and limbs. The largest redwoods, which are called redwood giants or redwood titans, are usually not the very tallest ones. In this way, they are rather like people. A football player is often bigger than a basketball player—more massive, that is. The basketball player is taller and more slender. So it is with redwoods. The tallest redwoods are often slender, and so they aren't the largest ones. Even so, the most massive redwoods (the redwood titans) are among the world's tallest trees anyway, and are more than thirty stories tall. Today, almost no trees of any species, anywhere, reach more than three hundred feet tall, except for redwoods. The main trunk of a redwood titan can be as much as thirty feet in diameter near its base.

Many people who are familiar with coast redwoods have seen them in the Muir Woods National Monument, in Marin County, just north of the Golden Gate Bridge. Muir Woods, which is visited by nearly a million people every year, is a tiny patch of virgin, primeval redwood forest, and it is like a small window that reveals a glimpse of the way much of Northern California looked in prehistoric times. Though the redwoods in Muir Woods are hauntingly beautiful trees, they are relatively small and are not very tall, at least for redwoods. The redwoods you can see in Muir Woods are nothing like the redwood titans that stand in the rain-forest valleys of the North Coast, closer to Oregon. These are the dreadnoughts of trees, the blue whales of the plant kingdom.

Nobody knows the ages of any of the living giant coast redwoods, because nobody has ever drilled into one of them in order to count its annual growth rings. Drilling into an old redwood would not reveal its age, anyway, because the oldest redwoods seem to be hollow; they don't have growth rings left in their centers to be counted. Botanists suspect that the oldest living redwoods may be somewhere between two thousand and three thousand years old—they seem to be roughly the age of the Parthenon.

THE ROAD BECAME THE CALIFORNIA COAST HIGHWAY, AND THE SILLETT brothers and Marwood Harris drove past Jedediah Smith Redwoods State Park, in Del Norte County. They didn't stop to look at the redwoods there. They went through Crescent City, a tired-looking town. They passed a Carl's Jr. fast-food restaurant, and a lumber mill, and bars and taverns, dark in daylight, where you could get a beer for a dollar and maybe get a fractured skull for nothing.

The redwood forests around Crescent City had been logged. The road went past stretches of open land covered with bare stumps, and past seas of young redwood trees growing on timber-company land, which looked like plantations of fuzzy Christmas trees. Here and there on the ridges were a few last stands of virgin, ancient redwoods, looming above everything else. They looked like Mohawk haircuts.

The road entered Del Norte Coast Redwoods State Park, and the highway was suddenly lined with extremely tall redwoods. Steve Sillett began thrashing around in the back of the Crypt. "Stop the car! I'm getting out."

Marwood pulled off to the side of the road. Steve squeezed out of the back seat and took off, running into the forest. Scott and Marwood waited in the car.

"What's he doing?"

"He's looking at the trees."

"Oh, God."

They rolled down the windows. "Steve! We're not there yet! Get back in the fricking car!"

TWENTY MILES FARTHER DOWN THE ROAD, THEY CAME TO PRAIRIE Creek Redwoods State Park. The park occupies a sliver of wrinkled terrain, eight miles long and four miles wide, lying along the Pacific Ocean on the northern edge of Humboldt County. The North Coast along those parts is covered with rain forests, and the forests are often hidden in clouds and fog. The beaches along the North Coast are made of gray sand, gnawed by waves the color of steel. The beaches rise into bluffs, which become the California Coast Ranges, a maze of ridges and steep, narrow valleys, clad with deep temperate rain forest. The forest is dominated by coast redwoods.

As they entered the park, Steve was hunched over, staring at a map. Marwood slowed to a crawl. Trucks whipped past them. Steve ordered Marwood to stop, and he pulled off the highway and rammed the Blue Vinyl Crypt into the underbrush, to get it out of sight. They were planning to camp in some wild spot among the redwoods, but it is illegal to camp in the redwood parks except in a few public campgrounds, and they didn't want the rangers to notice their car.

They put on their backpacks and hurried along a trail that went westward, climbing toward a ridge and the ocean, passing through a redwood forest. The trees had stony-gray bark. They looked like the columns of a ruined temple. The ground was made up of rotting redwood needles, and it was covered with sword ferns—tall, stiff ferns—growing chest high. Everywhere there were spatters of redwood sorrel—small, emerald-green plants with heart-shaped leaves.

The trail came to the crest of a ridge and dropped down into a valley that opened toward the ocean. As they went over the ridge, the sound of trucks on the highway faded away. A hush came over the world, and it grew dark. There was no sunlight at the bottom of the redwood forest, only a dim, gray-green glow, like the light at the bottom of the sea. The air grew sweet, and carried a tang of lemons. They became aware of a vast forest canopy spreading over their heads.

STEVE SILLETT MOVED OUT AHEAD ALONG THE TRAIL, AND MARWOOD Harris followed close behind him. Scott Sillett lingered, holding his binoculars in his hands and looking and listening for birds. It was so quiet in the redwood forest that he could hear the sound of his breathing. The trunks of the redwoods were grooved pedestals extending upward into hidden structures. He imagined that they were silent ru-ined towers of Middle-earth. Birds were moving in the canopy, but the birds were few and were quiet, for they don't sing in the autumn. He was hoping to see a varied thrush.

"Scott, we need to keep moving." Steve was standing up ahead, tapping his foot restlessly.

Scott watched his younger brother disappear down the trail. He thought that there was something fragile about Steve. Steve was a restless person, driven, passionate, intense, and he always seemed to be running out of time. He concealed his insecurity and sensitivity in a shell of prickliness and a weird sense of humor. Steve had a gloomy streak, and a tendency to be moody, to become angry and depressed, as if he had a hidden wound that oozed and could never heal. He also had an impulsive, generous nature, and his kindheartedness could get him into trouble.

The Sillett brothers had grown up sharing a bunk bed in a little bedroom in a duplex house with a neat yard in Harrisburg, Pennsyl-vania. Scott had claimed the top bunk, because he was older, and Steve had slept in the bottom. As children, they had invented a private language, which no one else could understand, and they still some-times spoke it with each other. It sounded like some kind of bizarre German. Their father, Terence B. Sillett, had majored in mathematics in college but had been unemployed for a number of years. Before he stopped working, he had been a salesman. He had sold insurance, real estate, and auto parts, and he had also worked as a chimneysweep, but he had developed spiritual longings. Eventually, he grew a beard and began meditating and spending a lot of time reading books on reincarnation and Hinduism and esoteric forms of Christianity. "I wanted to understand what the being called Christ was," Terence Sil-lett explained to me once.

The boys' mother, Julianna Sillett, discussed matters with Ter-ence, and they decided to switch roles. She became the full-time money-earner in the family. She was a registered nurse, and she got a job working in the labor-and-delivery room at a hospital in Harris-burg. Terence stayed at home during the day and took care of the boys and their sister, Liana, the youngest of the Sillett children. He did the grocery shopping, cooked meals for the children, made their school lunches, supervised their homework, and he tucked them into bed when their mother was working the night shift at the hospital. Alcohol became a steady companion of Terence Sillett.

In their little bedroom, the Sillett brothers found openings into other worlds. They became deeply involved with the books of J.R.R. Tolkien, especially *The Silmarillion* and Tolkien's *Unfinished Tales,* and with *The Chronicles of Thomas Covenant,* a series of fantasy novels by Stephen R. Donaldson. They spent days and weeks on end playing Dungeons & Dragons in their bedroom, sometimes staying up until one o'clock in the morning with the game. Terence would go upstairs to tell them to get to bed, and he would find his sons lying on the floor with a twenty-sided die between them, lost in a quest. The brothers' fantasy games involved friends of theirs, too. Scott's character was assassinated through treachery—some of his friends conspired to kill him—and after that Scott was reborn as a Dungeon Master.

For a while, Julianna didn't focus on the fact that Terence was drinking all the time. She had seen alcoholism in patients at the hospital, and she was shocked that she had been unable to recognize it in her husband. She told him that she would end their marriage if he didn't get himself into a hospital. He checked into the rehab unit of the Caron Foundation in Pennsylvania, and he stopped drinking. Back home, he continued his meditation and his spiritual researches. Steve broke off his relationship with his father for a time, and Scott's relationship with him was awkward and painful. Terence Sillett ended up becoming an addictions counselor with a private practice.

The Sillett family never had much money. When they went on vacation they camped out, or they visited Terence's parents, Charles and Helen Poe Sillett, who had a cabin in the mountains near Gettysburg. The boys' grandmother was known in the family as Poe. When Steve was eight and Scott was ten, Poe began taking them on bird-watching hikes in the forest around her cabin. Poe was a gruff, big-boned woman with flaming auburn hair and a scar across her face. She chain-smoked menthol cigarettes and was crusty and inscrutable. Squinting up into a tree through binoculars, she taught the boys how to identify birds. Poe couldn't always see them very well, but she could hear them perfectly, and she could tell their species by listening to their songs. She explained to the boys that a hermit thrush has a very different song from a wood thrush, and she taught them how to tell warblers apart by differences in their plumage. The song of the hermit thrush particularly moved Poe, and Steve noticed tears in her eyes as she listened to it, with a Salem stuck to her lip. Scott and Steve became birders as children, and they built up large life lists of birds they had seen. Poe had a passion for botany, too. She often used the scientific names of plants when she talked with the boys. To Poe, the buttercup was *Ranunculus*. Poe was not particularly fond of *Homo sapiens*.

She was a bit of a cheapskate. At Christmas and on birthdays, her main present to each of her grandchildren was typically a can of honey-roasted peanuts. She also gave them each five dollars, but she put the money in a savings account and told them she wouldn't let them touch it until they turned twenty-one. Poe believed that she was related to Edgar Allan Poe. Steve seemed to have inherited the writer's eyes: dark and big, looking out of the depths of a hidden mind. Poe offered to put a hundred dollars in the savings account of any grandchild of hers who would add the name Poe to his or her middle name, so that the famous name wouldn't vanish from the House of Sillett after she was gone. Steve and Scott thought that Poe's offer was a little cheesy, and they held out for more money, but Poe stood firm: "It will earn interest, and when you're twenty-one you'll get more than a hundred bucks," she told them, or something like that. They didn't add Poe to their names.

Steve finished high school in three years, with straight A's. He would come home on the school bus, hurry into the dining room, spread out his books on the table, and start doing his homework— often having forgotten to take off his coat. He went to a community college for a year and then applied to Reed, where he got a full scholarship.

When Scott turned twenty-one, he cashed in a lifetime of Poe dollars to buy a pair of bird-watching binoculars. They were Zeisses— really good. He had recently decided to become a scientist and study migrating songbirds. He had told Poe that she was the reason for his choice of a career, and she seemed pleased in a gruff way.

Where Steve was going was more of a mystery. As he had stared through binoculars up into the trees near Poe's cabin, searching for birds, he had found himself looking at the branches. They're huge flowering plants, he would say to himself. In some mysterious way that he would never be able to explain, Steve Sillett became involved with trees. There seemed to be another world in the trees. A forest was not what it seemed to be, but a web of life extending upward and out of sight. He would lose himself in the patterns of their branches. He wondered how old they were, and what lived in their tops. With Poe coaching him, he began to see how things in a forest are connected, sometimes invisibly, and how there is a logic to events as they unfold. Every year, as spring begins, birds arrive in a forest only after the insects hatch, because, before then, there is nothing for the birds to eat. The connections run through both space and time. Steve became sensitive to the movement of time in a forest. Time has a different quality in a forest, a different kind of flow. Time moves in circles, and events are linked, even if it's not obvious that they are linked. Events in a forest occur with precision in the flow of tree time, like the motions of an endless dance.

HIKING FAST, LEADING THE WAY ALONG THE TRAIL THROUGH THE redwoods that day in Northern California, Steve Sillett came to the banks of a west-running creek. He stopped and looked around. There was a grove of big trees there, near the creek. He wanted to see them more clearly. He went off the trail and began bushwhacking through undergrowth. Marwood went into the thickets after him.

Scott hurried along, and found that Steve was leading them into a tunnel of greenery that arched over the creek. The tunnel ended abruptly, opening out into a stand of big coast redwoods. Steve and Marwood took off their backpacks and explored the area, and eventually they ended up standing at the base of a large redwood, looking into its crown.

"What's your opinion, Tree Badger?" Marwood said to Steve in a low voice.

"I'm lusting for this tree."

Scott asked them what they intended to do. They didn't really answer him. They began walking in circles around the base of the redwood, staring up into it. The redwood was about three hundred feet tall—thirty stories tall—with a diameter of about fifteen feet near its base. The trunk was an enormous, furrowed cylinder, with no branches on its lower reaches. Far above the ground, a few whiffs of small branches popped out of the trunk, and then, higher up, a tangle of limbs emerged and wandered out of sight, buried in clouds of foliage. They couldn't see the top. It is doubtful that anyone had ever paid much attention to the tree, at least not since Europeans had come into the country. The land around Prairie Creek was originally inhabited by the Yurok people. As far as I know, the Yurok didn't give names to individual redwoods, although surely they were familiar with the giants. This nameless tree was a redwood giant.

"There's no way," Marwood said thoughtfully, gazing up the trunk.

"I see a way," Steve said.

Scott was standing next to them. "You guys aren't going to try to climb this tree, are you?" he said.

No answer. Steve took a few steps backward and then made a running leap at a small redwood standing next to Nameless. He grabbed a branch on that small tree that was eight feet above the ground, and suddenly he was standing on top of the branch—he had managed to swing himself up to it in a fluid movement. Then he began climbing from branch to branch, laddering his way up the little redwood. Marwood glanced at Scott with a distracted look on his face, dusted his hands together, and leaped up and began following Steve. The two Reed students were free-climbing the tree—climbing it without ropes or safety equipment.

"Hey. You guys? I really think this is not very smart," Scott called up after them. "You know what, we need to keep going. Are you listening? Steve? Marty? Steve?"

IN LESS THAN TEN MINUTES, STEVE SILLETT HAD GOTTEN CLOSE TO THE top of the small tree. The topmost portion of a tree is called its leader. The leader is a kind of finger that the tree uses to probe its way toward the sun.

Steve started climbing up along the leader, moving carefully and slowly over delicate branches. It narrowed to a pole that was thinner than his wrist, and it began swaying under his weight. Eventually he found himself standing seventy feet above the ground, balancing on a branch and holding the leader with one hand, having climbed as far as he could go, looking across at Nameless, the giant redwood.

A spray of branches frizzed out of a monstrous trunk across from him. The trunk was about twelve feet in diameter at this point, and it looked like a curving wall. A little branch stuck out of the wall, directly in front of him. He wanted to grab that branch. He edged closer to it, and the leader of the smaller tree began to bend under his weight.

There was a gap of empty space between the branches of the little redwood and Nameless. Marwood and Steve hadn't noticed the gap when they looked up from the ground. They had thought that the little tree's branches touched the big tree's branches, so they could walk across.

Steve studied the situation. If he could just reach out far enough he might be able to grab the little branch on Nameless. This is going to be a big extension, he thought. Keeping one hand wrapped around the leader, he reached his other hand out as far as possible. He ended up standing spread-eagled. He threw his weight outward. The little tree started bending closer to Nameless, but the target branch was still out of reach.

He was in a crux.

Marwood Harris clung to the branches below Steve Sillett, staring up at him through his gold-rimmed spectacles.

Steve paused. He looked up into the crown of Nameless. It was deep, explosive, mysterious. He looked seventy feet down.

Steve Sillett suffered from acrophobia, a fear of heights. It was a compulsive, uncontrollable fear of heights, which most people have to some degree and some people have to an extreme degree. His acrophobia wasn't overwhelming, but it could come on suddenly. Up in a high place, he would begin to feel dizzy and a sense of panic would creep into his soul, and his mind would fill with a horror of the void below. He would start whispering to himself, saying, Oh . . . oh, and he would begin to get a sensation of weightless acceleration, until he could actually feel his body falling down through space to death.

A distance of fifty feet above the ground is known to climbers as the redline. They hold it as a rule of thumb that if you fall fifty feet to hard ground you will very likely die. Indeed, an adult human can easily die after falling ten feet, if he lands on his head.

A person who is dropping in free fall through space often turns upside down and falls headfirst. This happens because, with most people, the upper half of the body is heavier than the lower half, and so the person tips over and plunges head-downward, like an arrow with a weighted point. In a headfirst landing after a fifty-foot fall, the shock crushes the skull and breaks the neck, destroying the brain and shearing the spinal cord off at the base of the skull. Instant death. No matter which way the victim lands, the impact normally breaks the victim's back, leaving him paralyzed (if he survives). If the person happens to land feetfirst, a shockwave travels up the legs, breaking them in many places, shattering the lower spine, and cutting the spinal cord. The lungs can collapse from the force of the impact, or they can be punctured by broken ribs, and the flattened or torn lungs can fill up with blood, causing the victim to drown in his blood. Major internal organs, including the liver, the kidneys, the spleen, and the bladder, as well as the aorta, can burst during the impact. If they split apart, they flood the body cavity with blood—catastrophic internal hemorrhage.

Steve Sillett was at about the height of the big top in a circus. He began to feel a bad sensation in his stomach, as if he were about to let it all go in his pants. Not good, since he was standing above Marwood. He tried to focus his mind on the problem. The gap was really not very large, he thought. He would have to let go of the small redwood and make a leap into the big tree, and catch a branch with his hands, like catching the bar of a trapeze. He had to jump high, or his body wouldn't clear the branches of the little redwood and he'd get tangled and would fall. If I was standing on the ground and I had to make this jump, I could do it, he thought. So if it's physically doable on the ground, why can't I do it up here? He tried to force his hand to just let go of the tree. Just let go.

DOWN ON THE GROUND, SCOTT SILLETT SAW THAT HIS BROTHER WAS getting ready to jump. "Oh, my God," he whispered. "Steve!" he cried out. "What the hell are you doing? Are you crazy, Steve?"

No reply.

"Listen to me! *If you fall, you will die.* Do you understand that concept?" Scott screamed. He tried to get Marwood's attention. "Marty, talk to him! Tell him not to do it! . . . You fucking idiots, we're miles from help! If you guys fall, I'm going to have to squeegee your corpses off the forest floor! I'm going to have to drag your shattered bodies out of here. . . ."

Marwood Harris wasn't really listening. This is just one of those hairy things, he said to himself.

"Goddammit!" Scott screamed at his brother. "Listen to me!"

The branch on the big redwood looked delicate. Steve couldn't really tell if it was strong enough to support his weight.

The lowest branches on the trunk of a redwood are called epicormic branches. Epicormic branches occur on many kinds of trees. They often arise out of scars left from broken branches, and they often come out in fan-shaped sprays called epi sprays. In redwoods, these branches are not solidly attached to the trunk, and can easily break or fall off the tree.

Steve Sillett didn't know what an epicormic branch was.

There was another problem, although Steve and Marwood didn't realize it. Hanging from the branch was a yellow jackets' nest the size of a cannonball; it wasn't visible from the small redwood.

"I can't watch you die, Steve!" Scott shouted. "I'm getting out of here! I'm going to look for birds."

Scott swept into the undergrowth. He was furious with his younger brother. He started hiking down along the creek, heading for the sea. He didn't go very far. How could he abandon Steve? He couldn't, so he waited by the creek in silence, feeling incredibly angry. The only thing he could hear was the sound of running water. He dreaded hearing a scream followed by a meaty thud.

In the top of the young redwood tree, Steve Sillett let go, and jumped into redwood space.

THE KINGDOM

AS STEVE SILLETT LEAPED OUT OF THE SMALL TREE, ITS TOP sprang upward. He felt gravity go to zero, and everything seemed to slow down. Feeling alert and calm, he watched the branch of the large redwood approach with drawn-out and perceivable slowness. The branch came into his hands. He closed his fingers around it.

There was a jerk. He found himself hanging from the branch by both hands, bouncing, with his feet kicking around in the air. He immediately began to travel along the branch, going hand over hand as he hung from it, trying to get himself over to the trunk. It wasn't a difficult move, it was just a monkey hang. Any kid could do it on a jungle gym. He didn't dare look down. He got himself over to the trunk, then he snapped his body upward and got one foot on the branch, and in one fluid movement he stood up on the branch, grabbed another branch above him and swung up to it, and stood balancing, motionless.

"It's doable," he said to Marwood. He was surprised at what had just happened, and his thoughts went: we're primates, and those opposable thumbs are awesome.

Marwood climbed into position and got ready. He thought, Well, if Steve can do it . . . He actually felt fairly confident that he could pull off the move. It was just that there was no safety rope to stop his fall if he made an error. Blow the move and lose your life. He got himself into focus and jumped.

THE NEXT THING STEVE KNEW, MARWOOD WAS HANGING BY HIS HANDS from the lowest branch of Nameless and screaming. The branch was shaking, and Marwood was writhing under it. He looked as if he were getting electric shocks. "Shit, there's bees!" he screamed. He was surrounded by darting yellow jacket wasps.

Marwood traveled hand over hand in a monkey hang, while the wasps stung him on the neck and hands.

For one bad moment, Steve thought that Marwood might get stung in the eyes, and would let go of the branch and fall to his death. But Marwood wasn't about to let go, no matter what. He got over to the trunk, got his feet planted in the bark, and hauled himself up onto the branch and stood on it, swearing loudly. Holding on to the bark with one hand, balancing himself on the branch, he swatted the yellow jackets off his neck and face, and then, still cursing, he began to climb upward at remarkable speed.

The approach of Marwood Harris in a swarm of wasps encouraged Steve Sillett to keep climbing higher. He went from branch to branch. Marwood followed below him. The yellow jackets faded away.

Farther up the trunk, Steve arrived at a blank zone, where there were no branches. He couldn't see any way to climb higher. Above him, extending twelve feet, was an empty, grooved surface of bark. At the top of the stretch of empty trunk there was a strong-looking branch.

He looked down. Mistake. He was ninety feet up. The fear of heights came over him in a sickening wave, worse than before. What am I doing here? he thought. I'm going to die. He began to climb upward along the surface of the bark, Spider-Man style, jamming his hands and feet into cracks in the bark. This climbing tactic is known to rock climbers as crack jamming, or crack climbing. To perform a

FRAGILE. Epicormic branches on a redwood. Like dog's hair, many of these branches are shed from the tree. *Drawing by Andrew Joslin.*

crack-climbing move, you jam a hand or a foot into a crack, get it stuck there, and then put your weight on the stuck part of your body and lunge upward with one hand, jamming it into a crack that's higher up.

With a series of crack-climbing moves, Steve climbed up the twelve feet of sheer redwood bark. He managed to get to the next branch, and he grabbed it, swung himself up onto that branch, and stood on it. He was somewhere in the lower tiers of Nameless, in the patchy zone of epicormic branches, where sprays of epi branches grow like fuzz on a redwood's trunk. He badgered Marwood, telling him it was easy, dude, and that he should have no problem getting up twelve feet of bare bark.

They both ended up standing on a big branch, at around 120 feet above the ground. They had arrived at the base of the crown of Nameless.

IN ITS FIRST TWENTY YEARS OF LIFE, A COAST REDWOOD CAN GROW from a seed into a tree that's fifty feet tall. In its next thousand years, it grows faster, adding mass at an accelerating rate. A redwood can go from a seed to a big tree in about six hundred years. Around age eight hundred, which is the end of its youth, it may reach its maximum height—its thirty-something-story height. Redwoods are extremely shade-tolerant. They can survive in dark places, at the bottom of a forest, in the deep shade of their elders, where few other trees would survive. A small redwood living in deep shade hardly grows at all, but it doesn't die; it goes into a kind of suspended animation. If it is hit by light, it grows with relentless speed.

As a redwood enters middle age, it typically loses its leader. Its top spire dies back. I suppose this is a little like a man going bald. The top spire turns into a bleached skeleton, and it falls off the tree. The bark of a redwood is usually a reddish-brown color, but as some redwoods get old their bark can turn stony gray. Even when its top falls off and its bark turns gray, a redwood is still growing and, in fact, may have hardly begun to grow. In this respect, it is not like a person at all.

A redwood reacts to the loss of its top by sending out new trunks. The new trunks appear in the crown, high in the tree, and they point at the sky like the fingers of an upraised hand. The new trunks grow straight up from larger limbs, rising vertically and traveling parallel to the main trunk. As the new trunks rise and extend themselves over centuries, they send out branches. These branches eventually spit out yet more trunks, and those trunks grow branches that send up more trunks, and so on. The tree is becoming a grove of redwoods in the air, containing redwoods of all sizes, from tiny to large. This aerial grove is connected to the ground through one main trunk. The whole structure is, of course, a single living thing.

With the passing of centuries, the extra trunks begin to touch one another here and there. The trunks and branches of a redwood can fuse and flow together like Silly Putty melting into itself. The bases of the extra trunks bloat out and become gnarled masses called buttresses. In the crowns of the largest redwoods, bridges of living redwood are flung horizontally from trunk to trunk and from limb to limb. This cross-links the crown and strengthens it, in much the same way that flying buttresses support the structure of a cathedral. As a redwood gets very old, its crown spreads out, until it can look like a thunderhead coming to a boil. Every five hundred years or so, a redwood may burn in a huge fire. There may be blackened chambers, holes, and rooms in the tree. These are fire caves. After it's been burned in a fire, even riddled with fire caves, a redwood can grow back. It drops off its charred buttresses and burned trunks, or it sends living wood flowing up around them, and it continues to send out new trunks, until the tree has become a three-dimensional maze in the air, with a complexity that comes close to defeating the human mind's ability to understand it.

By the measure of overall size—the volume of wood—the largest species of living tree on earth is not the coast redwood but the giant sequoia, a type of cypress that is closely related to the coast redwood. The giant sequoia occurs in sixty-seven small spots on the western slopes of the Sierra Nevada mountains in California. (The species name of the tree is *Sequoiadendron giganteum*.) The giant sequoia thrives in sunshine, and it gets moisture from melting snow. Giant sequoias (mountain trees) and coast redwoods (coastal fog-belt trees) are never found in the same forest.

The giant sequoia trees typically have thicker and more massive trunks than the existing coast redwoods, but they aren't as tall as coast redwoods. At present, the world's largest living thing is a giant sequoia named the General Sherman. The General Sherman Tree grows in Sequoia National Park, in the southern Sierra Nevada, and it is a tourist attraction. The General Sherman is twenty-seven feet in diameter at breast height. ("Breast height," four and a half feet above the ground, is the height at which tree experts measure the diameter of a trunk when they're calculating the size of a tree.) The General Sherman is 275 feet tall; it is a very tall tree. Even so, the General Sherman is more than a hundred feet shorter than the tallest coast redwoods. It contains around fifty-five thousand cubic feet of wood. If the wood in the General Sherman were to be cut into foot-wide planks an inch thick, and the planks were laid end to end, they would stretch for 125 miles.

Until recently, the coast redwood seems to have been not only the tallest but also the largest tree on earth, actually bigger than the giant sequoia. The largest redwoods were cut down in logging operations during the twentieth century. One of the larger redwoods was the Crannell Giant, which was felled in 1926 along the Little River, in Humboldt County. Records made when the Crannell Giant was cut up into cylinders show that it contained at least sixty-five to seventy thousand cubic feet of wood—it was a lot bigger than the General Sherman, and it was a lot taller, too. The coast redwood tree seems to be the largest and tallest individual living organism that has appeared in nature since the beginning of life on the planet. (There is something much bigger than a redwood, but it may not be strictly considered an individual. It is an edible fungus called the honey mushroom, or Armillaria, and the largest of those to have been discovered, so far, is a mass that lives mostly underground, putting up individual mushrooms here and there. The whole mass occupies a little over three square miles of the Blue Mountains in northeastern Oregon.)

STEVE SILLETT AND MARWOOD HARRIS BEGAN TO PENETRATE THE crown of Nameless. The branches were bigger and closer together, and there was foliage everywhere—above, below, and on all sides. At this point, they lost sight of the ground. They felt as if they were passing through a membrane and entering another world. Below them they could see nothing but opaque masses of greenery. Above them there was no sky—nothing but layer upon layer of foliage, like tents within tents. They had entered the deep redwood canopy.

The tree hadn't looked so big from the ground. Marwood could see Steve's sky-blue bandanna bobbing and weaving among the branches overhead. Steve was such a graceful climber, Marwood thought. Yet there were things about Steve Sillett that annoyed him. He worked out in the gym constantly, and Marwood thought he was a little narcissistic. He was incredibly good-looking, but he seemed to have no understanding of women, and he was barely removed from virginity. He complained endlessly to Marwood about how he couldn't find a girlfriend. His idea of making himself attractive to a woman was to lose his shirt: the shirt would come off, revealing a body that was like a statue of Apollo; but then he would start talking obsessively about moss and seed pods, things like that. It wasn't completely effective. Of course, he was still a teenager, a year younger than most college juniors, because he had skipped a year of high school. Reed College was an intense place, with high academic standards. It was clear that Steve was under a lot of pressure. To help blow off some of the pressure, Steve had started climbing trees on campus. Marwood and a few other friends had gotten into the habit of following him up into the trees, and they would chill in the treetops and talk, or take books with them and study. "If it wasn't for climbing trees and playing Dungeons & Dragons, I don't know how I could survive college," Steve said to Marwood one day. Steve Sillett was a science geek trapped inside the body of an Olympic athlete.

Steve was climbing twenty feet above Marwood, and he was concentrating on movement, keeping himself focused so that he wouldn't fall. At the same time, his senses were being almost overwhelmed by an impression of life all around. The crown of Nameless was loaded with clumps of ferns. The ferns were growing on the branches and down in crotches, where limb systems came together or where trunks grew out of limbs. There were hanging gardens of ferns and other plants. These plants of the air, he realized, had to be rooted in *something*.

He looked more closely. They were rooted in pockets of dirt. How had the dirt gotten up here, he wondered? He noticed that the branches of Nameless got larger and thicker higher up the trunk of the tree, farther away from the ground. This redwood structure is the opposite of most trees, in which the branches get smaller higher up in the tree. But the redwood forest grew increasingly complicated and more massive higher off the ground. He could only wonder how old the limbs were. If the tree was more than a thousand years old, then some of the larger limbs he was climbing on could have existed at the time of Columbus's discovery of the New World.

There were lacy things everywhere: flaky, gray-green, and brownish things, sort of like plants. They were lichens. Poe had taught Steve about lichens. Lichens are small organisms that often grow on bark and on rocks. A lichen (sounds like "liken") is a fungus growing in association with a species of alga or cyanobacterium, forming a single combined organism. Algae use chlorophyll to get energy from sunlight, as plants do, during photosynthesis. Cyanobacteria also collect light and perform photosynthesis. In a lichen, the fungus part of the organism is like a landlord who builds a house, and the alga or cyanobacterium is the tenant, who pays rent to the landlord. The landlord uses the rent to build a larger house for more tenants.

Steve had been interested in lichens since childhood. He found a kind of lichen growing on Nameless that Poe had shown him as a boy, when they were out catching toads and snakes near her cabin. It was British soldier lichen, which forms little stalks with red caps on them, supposedly like British soldiers. Poe had taught him its Latin name, *Cladonia,* saying it in her smoke-wrecked voice.

There were many other kinds of lichens in the tree, in all manner of shapes—drippy, frizzy, frilly, powdery, crusty, stringy, and hairy lichens. He began picking pieces of lichen off the branches and tucking them into his shirt pockets. There were mosses everywhere, as well, of different shapes and shades of green, and there were other kinds of plants growing on Nameless, too.

A plant that grows on another plant is called an epiphyte. An epiphyte is an aerial plant, and typically grows on the branches of a tree. It does not have its roots in the ground. Epiphytes feed on the moisture and nutrients in trees. Steve felt as if he had gone inside the body of a living organism, with uncounted other organisms living inside it. The fear of heights finally left him, and was replaced by a feeling of almost indescribable wonder.

At the time that Steve Sillett and Marwood Harris made the first ascent of Nameless, there was a general belief among biologists that the redwood forest canopy was what they called a redwood desert. That is, the redwood canopy was believed to be essentially empty of life other than the branches of redwood trees. In any case, biologists regarded coast redwood trees as unreachable towers, remote and bare. Steve Sillett encountered something quite different. He found what amounted to coral reefs in the air.

MOVING UPWARD STEADILY THROUGH THE LABYRINTH, STEVE AND Marwood reached the top of Nameless an hour after they had jumped into it. As they neared the upper surface of the redwood canopy, they saw that sunlight was shining through a Gothic lacework of branches, illuminating masses of plants, which glowed with varied shades of green. Steve, leading the climb, broke out into a lost glade in the sky.

The top of Nameless had been sheared away in a storm that occurred many centuries earlier, and the tree had reacted by driving a radiance of branches spreading horizontally in all directions away from the broken trunk, like spokes coming from the hub of a wagon wheel. Those branches had sprouted vertical trunks, like spikes on a crown. A forest of small redwoods had sprung out of the top of Nameless—Nameless Wood.

In the center of Nameless Wood, there was a big, rotting stump, several feet across. It was the broken-off main top of the tree. Billows of huckleberry bushes grew out of the stump in beachball-size puffs, and they were full of ripe huckleberries. Steve sat down beside a huckleberry bush. He couldn't believe what he was seeing. It was sunny at the top of the tree, warm and autumnal. He began plucking and eating the huckleberries. They were sweet and slightly tart, and they dripped with juice. There were damp spots in the top of Nameless. It was as if there were tiny springs in Nameless Wood, and the huckleberry bushes were sending roots down into the moisture.

Marwood Harris's head appeared through a gap in the platform of branches. He looked around. "Holy shit," he said. The branches and trunks of Nameless Wood grew so close together that Steve and

Marwood could climb around on them as if they were climbing in a jungle gym, without fear of falling.

Marwood explored the top of Nameless, and then he sat down on a branch beside Steve and stretched out, warming his body in the sunlight. The spiritual weight of the place seemed immense. It was as if he were waking up from a sleep, as if his life up to then had been a dream, and this was real. He felt as if he had left time behind. He felt as if he were leaving Marwood Harris behind, too. What seemed real to him was the trees. The redwoods were overwhelming in their reality. He could see nothing except the California Coast Ranges, running off to the north, south, and east, ridges and canyons blanketed with redwoods, and then the Pacific Ocean, stretching westward into infinities of blue. Redwoods and the sea. They spent a long time resting in the top of Nameless, saying little to each other. They had discovered a lost world above California, and it was unexplored. The sun began to drop down over the ocean, and the sky deepened in color.

WHAT WAS DISMAYING WAS THAT THEY WERE ONLY HALFWAY THROUGH the climb. They still had to get down, and this was the part that really frightened Marwood Harris. Climbing down is scarier than climbing up. They began to descend, very slowly. As they went down, they came to places where they had to hang from a branch by their hands and let go, drop through the air down to the next branch, and grab it for dear life when they landed on it. More than once they nearly slipped to their deaths. And then they had to make the leap back across the terrifying gap into the small redwood. The wasps were stirred up and attacked them again. Eventually Scott Sillett heard their voices in the little redwood, and they made it to the ground; their faces were blotchy with stings, but they were alive.

That night, the three young men pitched their tents in a redwood grove not far from the sea. They ate canned spaghetti for dinner. As he listened to Steve talk about the climb, Scott began to worry about his brother. Steve was restless, like their father, and he was driven by passions that caused him to risk his life. Scott could see the beginning of a story involving his brother, but he couldn't see the end. The question, to Scott, was whether his brother would survive.

Drawing by Andrew Joslin.

ISLAND IN THE LAKE

FROM A HUMAN PERSPECTIVE, TIME IN THE REDWOOD FOREST moves slowly. Seasons pass, and the redwoods seem unchanged to the human eye. They are evergreen, and keep their foliage year-round. Botanists refer to the needles on conifers as leaves. Each leaf of a redwood lasts about seven years before it falls off, in the autumn, when redwoods lose their oldest needles. A redwood may increase its diameter by only one millimeter a year, or it may add almost nothing to its diameter, only the thickness of a single layer of cells, if it is encountering conditions that discourage growth. Even so, redwoods are forever in motion, extending upward into the canopy and fighting to fill space. Time in the redwoods, from the human perspective, can move so slowly that it seems to spread out until it has the quality of a vista. Placed against the backdrop of redwood time, a human lifetime shrinks into a compressed flicker, and the past, present, and future seem to run together and vanish.

One summer day near the end of July 1980, (a blink ago in redwood time), a girl named Marie Antoine was exploring a tree. The tree was a balsam fir, and it grew on Treaty Island, in Lake of the

Woods, Ontario, Canada. Treaty Island is two miles long, and its shoreline is scalloped into coves and points. The island is covered with white pines and red pines that grow on rocky bluffs, and its interior rises to meadows dusted with wildflowers. Marie was four, almost five—her birthday would be in just a month—and she was small for her age. She had long brown hair and a delicate face, and her eyes were hazel, but the edges of the irises were blue. The balsam fir that she was climbing was green and bright and fragrant. It grew beside a mansion that she knew as the Big Blue House, a rambling old Edwardian structure with twenty-two rooms and three round turrets springing out of its wings, where her grandparents lived. Almost assuredly, nobody had ever climbed the tree before. Marie was its first explorer.

She was determined to climb high enough so that she could look down on the turrets of the Big Blue House. Maybe she could even get up on the roof and climb around and find things, or climb in a window and surprise her grandparents. She climbed upward, going from branch to branch. She wasn't the least bit afraid. The trunk narrowed, and it got thin. As she got higher in the tree, she could smell balsam blended with a summery smell of lake water hanging in the air. She loved the smell of Lake of the Woods in the summertime. Often, she heard the sounds of motorboats in the distance, people crossing the lake from one island to another or going to town.

The Big Blue House had secret passageways in it. One secret passage started in the back of Marie's grandparents' closet in their bedroom and went up a ladder into the East Turret. It was haunted by the ghost of Mrs. Lystico. Marie was absolutely terrified of the ghost of Mrs. Lystico. She had been a rich old lady who had once lived in the Big Blue House. They said that something really bad had happened to her, something so terrible that she decided to kill herself. Marie's cousin Renée had once seen Mrs. Lystico's ghost weeping. Marie couldn't bear to look inside her grandparents' closet, and she hoped that she would never get a glimpse of Mrs. Lystico, walking around in the turret and crying about how she was dead.

As Marie Antoine climbed higher into the balsam fir she could see her father working in the garden far below, tending his peonies and delphiniums. Ronald Antoine was a tall, handsome man, with brown hair and an easy smile, and he had long sideburns and short hair, and a lean, kindly face. Marie knew that she'd better not call out to him or make any noise: he might tell her to get down. She climbed quietly, keeping her mouth shut.

Finally, she was about forty feet above the ground, higher than the roof. Way up. The tree was getting hard to climb. Marie found a nice, comfortable branch, and she sat down on it. She felt safe up in the tree.

At the place where she was sitting, the bark was bumpy with sap bubbles. She broke off a twig and poked it into one of the bubbles to see what would happen. A jet of balsam sap squirted out, and droplets stuck to her face. She liked that, and it smelled good. She started popping more bubbles with the end of the twig. It was very satisfying, but the sap squirted over her clothes and also into her hair. Finally, she had popped all the sap bubbles at that place, and she got bored. She had done everything interesting in this tree, so she climbed down, going carefully from branch to branch. Nobody had noticed her there.

Marie Antoine lived in a small house called the Cottage, which was down the hill from the Big Blue House. The Cottage was perched on a cliff overlooking the lake, and a path led down from the Cottage to a dock and a boathouse in a cove.

Her hair was sticky. She ran along the path through the trees to the Cottage. She went upstairs to her bedroom and got a comb. She tried to comb the sap from her hair, but it didn't work. The comb got stuck. It wouldn't come out of her hair, no matter what she did.

Marie's mother, Elizabeth Antoine, was sitting in a chair in the living room of the Cottage, by a plate-glass window, looking out over the lake. Marie's mother was a beautiful woman, with a soft, rounded face and dark, rich hair. There was a tea table beside her chair, with a book on it, and she had been reading quietly in her chair. She burst out laughing when she saw Marie with a comb stuck in her hair. "Oh no, Marie, you messed up," she said. She had a bright laugh, and her voice sounded like Kentucky, not like Canada.

The comb was really stuck. "We're going to have to cut off all your hair," Elizabeth said. She felt that this was something Marie's grandmother should see, and she needed a little help with it, too. She got up and walked slowly over to the telephone and called the Big Blue House. Pretty soon Marie's grandmother came by. She got a stool and put it next to Elizabeth. Marie sat on the stool so that her mother could reach over and touch her. It gave her a close, nice feeling as her mother cut off every bit of her hair. Marie ended up looking like a boy, but she didn't care. Her mother said that her hair would grow back, anyway, so never mind. Not long afterward, Marie's mother became gravely ill.

ELIZABETH WOULD SAY, "PLEASE DON'T CLIMB UP ON ME, MARIE, because it does hurt me a little when you climb on me." She explained to Marie that her bones might break if Marie got up on her lap. She sneezed once and broke two ribs. Around that time, Elizabeth needed to have two canes to help her walk. Marie's father placed the canes on either side of his wife's chair so that she'd have them within reach if she wanted to stand up.

Elizabeth's toes were strong enough. They weren't going to break—or if they did break, it was only her toes, so who cared. She kicked off her slippers for Marie. The toes were people. Each toe person had a different name and a different personality. They all talked in squeaky Kentucky voices, and they made Marie laugh. She got so excited by the toe people that she ran up to her bedroom and got her dolls and brought them downstairs so that the dolls could talk to her mother's toes. The dolls played and laughed uproariously with the toe people and everybody had an incredibly good time, including Marie's mother, all wrapped up in a blanket in her cozy chair, hardly moving, with just her toes sticking out and wiggling around and her bright eyes and a huge grin on her face. Elizabeth loved to tell stories, too, while she was all wrapped up in her blanket. Some of them were funny and dirty.

Ronald Antoine had once been a high school teacher in Southern California, where he and Elizabeth met. He had taught just about every subject: art, mathematics, science, history, English. Ronald and

Elizabeth had fallen in love, and eloped, because they didn't want anyone to make a fuss over their wedding.

Ronald's parents had owned and run the Big Blue House as a hotel in the 1950s (it was a landmark on Lake of the Woods), but gradually they had allowed it to become a summer house, just for themselves, and it was getting a little seedy and run-down. Ronald, when he was a schoolteacher, had bought a house in Malibu, and he found that he could sell it for a lot of money. ("Dad bought his house before Malibu was Malibu," Marie said to me once.) Meanwhile, Elizabeth had worked for the reclusive billionaire Howard Hughes, and when she left that job she had come away with some retirement money—Hughes was generous with his staff. Between the two of them, they had ended up with just enough money to live a very simple existence on Treaty Island. Ronald grew a lot of the food they ate, and they didn't own a car—Ronald disliked driving cars. They built the Cottage. They were determined to achieve a simple happiness on the island, and to spend time with their children, raising them in love and security. They began taking in foster children. These were Ojibwa kids, most of them toddlers, and they were always bumping around the Cottage and the meadow. One of them was a little Ojibwa girl named Bella. Ronald and Elizabeth adopted Bella, and she became Marie's younger sister.

MARIE ANTOINE WAS FOUR YEARS OLD WHEN HER MOTHER LEARNED that she had bone cancer. At the time of the diagnosis, the disease was far advanced, and her case was considered terminal. Elizabeth's doctor gave her about three months to live.

Elizabeth didn't think that Marie would remember much of her if she died in three months. Four-year-old children don't keep many memories. In spite of the pain of the bone cancer, which at times caused her to long for death, Elizabeth decided to try to keep herself alive using her strength of will. She believed that a person can stave off or delay death for a certain space of time, a period of grace, even when the body needs or wants to die and the mind aches for a release. This period of grace could be attained by self-discipline and love—her feeling for Ronald, Bella, and Marie. The mind could order the lungs to keep breathing and the heart to beat, no matter what kind of pain there was. Marie was just arriving at an age when memories could be formed and become permanent, and memory was the only gift that Elizabeth could offer to her daughters for the years to come.

THE START OF THE HUNT

THE FOREST CANOPIES OF THE EARTH ARE REALMS OF UNFATH-
omed nature, and they are vanishing. The earth's forests are
being logged off, burned away, turned into patches, and reduced to small fragments. Like other natural systems on the planet,
the forests are very likely being altered by global warming brought on
by the human production of carbon dioxide and other greenhouse
gases that are being released into the atmosphere. We know very little
about the forests or about what is happening to them, as little, possibly, as we know about the oceans that surround the continents. We do
know that whatever happens to the great systems of nature will also
be what happens to us.

The species that live in forest canopies are largely unknown. Biologists say that they are undescribed. A species that is undescribed is one
that has never been given a name or been identified according to what
it looks like, where it lives, and how it fits into the classification system
of life—the taxonomy into which all living things can be placed, from
bacteria to elephants. The forest canopies of the earth are believed to

hold roughly half of all species in nature. No one knows, exactly, because no one has a clear idea of how many species actually exist on the earth. There may be ten million different species on the earth, or a hundred million species. The forest canopy is the earth's secret ocean, and it is inhabited by many living things that don't have names, and are vanishing before they have even been seen by human eyes.

THE AMERICAN GOLD RUSH BROUGHT A STREAM OF SETTLERS UP along the coast of Northern California. At that time, around 1850, the redwood forest amounted to roughly two million acres of virgin, old-growth forest, stretching from Big Sur to a corner of southern Oregon. Old-growth forest is a forest in which the trees have been growing undisturbed for long periods of time. Old-growth forest has a high amount of biodiversity—a great variety of different kinds of plants and animals live there, in complicated niches and habitats, interacting in complicated ways.

Humans have lived in California for at least eleven thousand years, and they have long shaped its landscape. The mountains along the northern coast of California are covered with rain forest, yet the forest opens unexpectedly into beautiful stretches of grassland and prairies that run up the hillsides and over the ridges. Many of these grasslands are not natural, but are human works of great antiquity. The Yurok people, and other Native American groups living along the coast, traditionally burned the prairies in order to keep the redwoods at bay. There were oak trees growing in the prairies, and they provided acorns, which were a large part of the Indians' diet; the prairies also supported deer, which were a source of meat. The redwood forest provided logs for dugout boats and planks for houses, but it didn't supply much food. The Yurok made river canoes and large, oceangoing canoes out of redwoods, and sold them to other peoples, and together they exploited the redwoods as a source of trade and wealth.

When the European settlers arrived, they began cutting down the redwoods with axes and handsaws, using the wood to make barns, houses, fences, and railroad ties. In the late nineteenth century, the introduction of steam-powered logging machinery, followed, in the twentieth century, by gasoline-powered machinery and chainsaws, vastly increased the speed of logging along the northern coast of California, and the old-growth redwood forests began to disappear.

In 1918, a group of citizens founded a nonprofit conservation organization called the Save-the-Redwoods League. Some of the members of the group were wealthy, and they also had access to the money of other wealthy people, including John D. Rockefeller, Jr. They used the money to help buy up tracts of virgin redwood forest. Over the years, the Save-the-Redwoods League bought a total of about 170,000 acres and donated the lands to the California state redwood parks and to Redwood National Park, where the ancient redwoods remain. Most of the rest of the redwood forest ended up being owned by timber companies. As a rule, they carried out clear-cutting operations, in which no tree of any monetary worth was left standing. In all, close to 96 percent of the primeval redwood forest was cut down. What is left of the virgin redwood forest is like a few fragments of stained glass from a rose window in a cathedral after the rest of the window has been smashed and swept away.

A redwood is a tough tree, however, and when the tree is burned or sheared off at its base it has the ability to send up new sprouts from its root system. The root sprouts rise up in a circle around the stump. In the fullness of time, the root sprouts can become a circle of redwood trees, which is called a fairy ring. If all the trees in the ring sprouted from one stump, the ring is essentially a single organism. The DNA of all the redwoods in such a fairy ring is the same— in other words, the trees in the ring are clones, joined through their roots. A redwood fairy ring that has grown old and vast, and has fallen partly into ruin, is known as a cathedral.

Timber companies manage their forests for high-volume production of timber. The redwoods on timber-company lands are being cut on a regular schedule, every fifty years or so. Many of the trees are actually fairy rings coming out of old roots, and the roots may be up to thousands of years old. The fairy rings on timber lands are getting cut down repeatedly, and the old root masses under them may be fading away and losing strength, or maybe not. They may be fine—no one knows.

The largest and tallest redwoods often occur in small floodplains at the bottoms of valleys called alluvial flats. The soil in an alluvial flat is rich and moist. In an alluvial flat, a redwood is protected from winds, and it can drink unlimited amounts of water and grow exceptionally tall and large. The alluvial flats in the valleys of the North Coast were the core zones of the greatest redwoods. Loggers moved up along the valleys, laying railroad tracks and cutting the flats along the creeks, and the most magnificent groves were largely wiped out. But not entirely.

IN THE SUMMER OF 1963, A WRITER AND NATURALIST NAMED PAUL A. Zahl led a National Geographic Society expedition into the remains of the ancient redwood forest of Northern California. His purpose was to write an article about it for *National Geographic* magazine. Paul Zahl was a chunky man with dark-framed eyeglasses and a thoughtful, plainspoken manner, and he wore flannel shirts when he was tramping around in forests.

Melville Bell Grosvenor, the head of the National Geographic Society, had begun to believe that a national park should be established in order to preserve the redwood forests. Grosvenor seems to have thought that if some previously unknown redwoods of world-record height could be found growing on timber-company land, it would help strengthen the case for federal protection of the redwoods.

When Paul Zahl left for California, he spoke with Melville Bell Grosvenor. "Keep your eyes open," Grosvenor said to Zahl. "It would be wonderful to find a record-breaker."

Zahl brought along his wife, Eda, and their two children. They got a station wagon and drove around the redwood state parks, and eventually the family ended up staying in a motel near Redwood Creek, a narrow, fast-flowing river that runs through a canyon among steep mountains in Humboldt County. The area is covered with rain forest. At the time, the mountains and valleys around Redwood Creek were private land, owned by timber companies, which had been pushing roads into the terrain and clear-cutting it. The roads were coming in along ridge lines and moving downslope toward Redwood Creek. Many of the deepest parts of the canyon and the surrounding valleys were full of virgin redwoods growing on alluvial flats.

No roads or trails led up along Redwood Creek or into the canyons above it, which were virtually impenetrable, walled with vegetation. These rain-forest canyons are sometimes referred to as notch valleys. Some of the notch valleys that drained into Redwood Creek had probably not been visited by humans since Indian times, for not even timber-company men, searching for stands of large redwoods, had gotten into all parts of the region. Travel through the ancient redwood rain forest of the California Coast Ranges can be nightmarishly difficult. Not even the Indians had found it easy, it seems—they had stayed mostly along the bigger creeks.

The summer passed, and Paul Zahl flew back east for the start of school, but later he returned to California, determined to find a record-breaking tree. He started his explorations by going on a bush-whacking hike up along the lower parts of Redwood Creek. He carried with him an Abney level, a disk-shaped device that contains a mini-telescope and is used for measuring an angle to the top of some tall object such as a tree. Using an Abney level, you can estimate the height of the tree.

Redwood rain forest is choked with fallen logs ten or twelve feet across, vast standing trunks, and thick vegetation, and the land can be impossibly steep. Paul Zahl found that the easiest way to move was to walk along the creek or wade in the water. He measured redwoods here and there as he went along, but he didn't find any exceptionally tall ones.

He learned about the existence of a logging road that ran from a ridge overlooking the middle section of Redwood Creek. There, in October, he caught sight of a cluster of very tall-looking redwoods growing on a flat inside a bend of Redwood Creek. They looked like rockets taking off.

Zahl spent days slogging around in deep underbrush along the creek, trying to determine which trees there were the tallest. Eventually, he identified an extremely tall redwood that seemed to spike its way above the surrounding forest. It was growing less than a hundred yards from the river. He took some lines and angles on it, and got a preliminary estimate of 370 feet. The tree seemed to be by far the world's tallest.

Zahl telephoned National Geographic's headquarters in Washington, D.C. The effect was electrifying. Within days, Melville Bell Grosvenor and a team of executives from the National Geographic Society flew to Northern California. They were blown away when they saw the tree, and judging from the article that Grosvenor himself wrote in *National Geographic* it seems clear that his first sight of the world's tallest tree was one of the most exciting moments of his life. The Society engaged the services of a team of professional surveyors to measure the tree's exact height. This proved to be a very difficult task. After days of work, the surveyors finally reached a consensus that it was 367.8 feet tall—it was indeed the tallest known tree on earth. (Botanists measure trees in tenths of feet, rather than inches, or in meters.)

The naming of this tree was determined by politics. Its discovery had been sponsored and paid for by the National Geographic Society, whose top executives decided to name it the Libbey Tree, in honor of Howard A. Libbey, the president of the Arcata Redwood Company, which owned the land where the tree grew. Mr. Libbey had promised, enthusiastically it seems, never to cut down the trees in the grove (there were plenty of other redwoods in the area to cut). Later, the Libbey Tree came to be called, simply, the Tall Tree. The grove in which it stood was named the Tall Trees Grove. In 1968, Redwood National Park was established by an act of Congress, and the heart of the park was the Tall Trees Grove, on Redwood Creek. The Tall Tree was considered to be one of the wonders of nature in North America. *National Geographic* declared that it was "the Mt. Everest of All Living Things."

UNEXPLORED

ONE DAY IN THE MIDDLE OF OCTOBER 1987, A COLLEGE student named Michael Waring Taylor found himself on a shuttle bus full of tourists in Redwood National Park. The bus was winding its way down a dirt road, heading for the Tall Trees Grove. This was, it seems, during the same week that the Blue Vinyl Crypt was parked about fifteen miles north of there, in the underbrush at Prairie Creek Redwoods State Park, though Taylor had never heard of Steve Sillett or his companions. Michael Taylor was a junior at Humboldt State University, in Arcata, a town on the North Coast. He was a slender man, twenty years old, with a tall, wiry frame and dark-blond hair, and a pleasant, attractive face. Taylor's eyes were a peculiar blue-green color.

The bus stopped at a pullout, and a park ranger began helping the tourists to get off. Everyone in the group was older than Taylor, and many of them were retired. There was a cooler full of sodas for the tourists. Taylor picked out a root beer and began walking with the group along a trail that leads down to the Tall Trees Grove. He walked more slowly than the old people, looking around, sipping his root beer. The trail leveled off and came to the flats near the creek, and entered the Tall Trees Grove. The redwoods there were enormous. The air was heavy with the sharp smell of California bay laurel trees, which grew around the ankles of the redwoods. The tourists wound their way through the trees, stopping before a plaque in front of the Tall Tree. Taylor tipped his head back and let his eyes run upward, savoring the impossibly long shaft of the trunk, the sprays of branches as they flung themselves at the sun. His eyes were sensitive to bright light, and when he stared into the tops of the redwoods they began to water.

Michael Taylor suffered from an intense fear of heights. He didn't like to go into tall buildings. If he got too near a window on a high floor, his acrophobia would kick in and he would get an urge to throw himself out the window. It was as if some demon inside him were whispering, Jump, just do it—it's gonna be interesting. At those moments, he experienced an alarming sense of curiosity about what it would feel like to actually sail down through the air to his death.

Taylor moved away from the tour group. He couldn't stand lectures. He went down a slope, pushed through some bracken fern and river willows, and came out on a gravel bar on Redwood Creek. The water was low, crystal-clear, sliding over a bed of gray and blue stones. He took off his boots and waded barefoot across the creek. It was icy cold. He climbed out on a gravel bar on the west side of the creek. He was hoping to get a clearer look at the world's tallest tree.

Taylor became obsessed with redwood trees when he was a child. When he was about ten years old, his mother took him camping in Big Sur, where he saw his first redwoods, and the trees had gotten into him somehow. Why would a child get so whacked-out on redwoods? He couldn't say. And it was getting worse and worse, this redwood thing. He felt that he needed to go into some kind of redwood detox.

MICHAEL TAYLOR IS THE OLDEST SON OF JAMES SEARCY TAYLOR, THE chairman and CEO of American Capital Group, a private real-estate investment company headquartered in Santa Barbara. American Capital develops and owns shopping centers and office buildings throughout the United States. The company also owns some of the land

beneath the California stores of Costco Wholesale Corp., the huge re-
tailing chain that competes with Wal-Mart. Jim Taylor is a stocky,
athletic man with youthful looks and a pleasant, sociable manner, and
he has a wide circle of friends. In addition to his interest in American
Capital, Taylor is also part owner of a major-league baseball team,
the Oakland A's, and of golf residential communities and a ski resort
in Montana. He is passionate about golf and is a skilled fly fisherman,
and he collects and reads rare books. Like many people who have
built their wealth in real estate, Taylor's net worth seems a little hard
to figure out. He describes himself as "relatively well-to-do." When I
asked him if he was a demi-billionaire, he got a little testy. "Me, demi-
billionaire? Ha! It would be nice to have half a billion dollars, but I
don't. Maybe I will someday. What's important to me is the quality of
my life, especially my personal relationships with my family and
friends."

When Jim Taylor was seventeen and had just graduated from high
school, he had a summer romance with a sixteen-year-old high school
junior named Cassie Jane Johnson. She became pregnant with Michael,
and she and Jim got married so that the baby would not be born out
of wedlock. Their marriage seemed troubled from the start. In 1969,
three years after Michael was born, Jim and Cassie Jane had a second
child, Jeb, and Jim soon moved his family to Cambridge, Massachu-
setts, where he attended the Harvard Business School. At Harvard, he
joined in protests against the Vietnam War and the American bombing campaign in Cambodia. He was also a leading organizer of the
first Earth Day at Harvard. Eventually, he took a leave of absence
from Harvard (he would never graduate) and he returned to California with plans to become a forest ranger.

"I was quite certain that I didn't want anything to do with business, but I figured with a Harvard business education I could at least
get a job as a ranger," Taylor explained. He discovered, however, that
he wasn't qualified for a ranger's job, and eventually he went to work
as a partner at a real-estate firm in Santa Barbara. He later bought out
his partners, and the firm became American Capital. Eventually, Taylor got to know Jeff Brotman and Jim Sinegal, the founders of Costco,
which was then a regional discount retail chain with headquarters in
Seattle. The company wanted to expand into California. Taylor got

involved with the process, buying and managing parcels of real estate for Costco to build stores on. He helped Costco conquer California, where the chain grew spectacularly, as did the value of Jim Taylor's business.

Cassie Jane became a nurse. Their marriage fell apart, and they were divorced in 1972, when Michael was six and Jeb was three. Jeb ended up living with his mother in a little house in the hills above Santa Barbara. She married an artist named Norm Hendry, who made pottery, and he and Cassie began working together as potters. "My mother shopped for clothes at rummage sales, and her idea of entertaining was to go for a hike with some friends and then have them over for supper," Jeb Taylor says. Michael went to live with his father.

Jim Taylor purchased an estate in Hope Ranch, an exclusive community overlooking the ocean in Santa Barbara. The estate had a guesthouse and a main house. Michael lived in the main house, and he was often there by himself. When Jim wasn't away on business, he was preoccupied with entertaining friends. He threw big parties, owned half a dozen fast cars, and dated younger women. He was still in his twenties when Michael was a boy. "I was an immature father, and it was difficult for Michael," Jim Taylor told me. The main house was set into the hillside, and it had a swimming pool. There was also a fish pond, stocked with koi—colorful Japanese carp. The koi pond extended inside the house and under the main stairs. There was a live-in cook. Michael could tell the cook what he wanted to eat, and the cook would also run errands for him. On weekends, Jeb came to stay, and the brothers swam together in the swimming pool and played Ping-Pong. When Michael visited Jeb, their mother and Norm would take them on backpacking trips.

Jim enjoyed his weath, but he didn't want his children to develop a taste for money or a dependence on him. He wanted them to have good values and to cut their own paths in life. "Dad was cautious about giving us expensive presents," Jeb said. "Sometimes he'd give us twenty bucks to go to the video arcade and he'd never ask for it back, and other times he'd give us five bucks and make it clear that he pretty much expected us to pay him back."

Michael got heavily into Dungeons & Dragons, and he played it endlessly in his room by himself. He kept the window shades drawn, because he didn't like bright sunlight. He stayed away from the beach and started playing video games. His stepfather, Norm, the ceramicist, sometimes gave Michael odd tools and objects, and Michael piled the things in his room. He would wire stuff together, hoping to create special effects. One day he built an electric-shock device out of a car starter and a car battery. He took it over to his mother's house and hot-wired the metal latch to Jeb's bedroom door. Cassie Jane put her hand on the latch and got a pulse of electricity that would have started a Buick. She screamed, but fortunately she wasn't hurt in any serious way. Michael got a reputation in the family for being brilliant, eccentric, clever with gadgets, and hard to figure out. "In our family we've always been worried about Michael. Always," Jeb said.

Jim Taylor started backing golfers on the PGA tournament circuit. He and his golf-pro friends would fly to tournaments together, and sometimes he took Michael along. The golf pros gave Michael tips, and he became an excellent golfer. As Jim's business flourished, he bought a private jet, a Cessna Citation, which had two full-time pilots. He used the jet in place of a family car, to take weekend trips and family vacations around California and across the West. The family called the jet Triple Five Charlie Charlie, after the letters on its tail. Jim would take the boys to Lake Tahoe for a few days of skiing, or would fly up to Monterey with Michael so that the two of them could play a round of golf at Pebble Beach Golf Links. Occasionally, they took the family jet down to Los Angeles—a twenty-minute hop—to have lunch with some of Jim's Hollywood friends, who included the television producer Norman Lear. When they went to L.A. Lakers basketball games, they sat on the floor next to the team—Jim was friends with the Lakers' doctor. They took the jet to Super Bowl XIV, when the L.A. Rams played the Pittsburgh Steelers. Jim was friends with Don Klosterman, the general manager of the Rams, and they chewed the fat with him before the game. When they went to Alaska, Jim would hire a float plane to drop them on a wilderness river for a few days of fly-fishing.

Michael Taylor was intensely happy when he was flying with his father. They were unusually close in age for a father and son, and they got along easily. Michael developed a powerful drive in golf, he became extremely skilled at casting a dry fly on moving water, and he felt at home camping in deep wilderness, having learned these things with his father. "What a blast, flying around the country to meet pro golfers with my dad," Michael said to me. "I had a great childhood."

His social situation in high school wasn't as good. "I got bullied a lot," Michael said. He was quiet, and some of the kids called him a dork. "Michael didn't care about fashionable clothes; in fact, he didn't even seem to know what fashion was," Jeb said. Michael didn't drink, and he hated parties. His grades stank.

"From the moment Michael was born, he insisted on doing things his way," Jim Taylor said of his son. "I think of Michael as a little bit of an enigma. I see him as this remarkably talented human being in the body of someone who has a surprising lack of concern for the things that most of us regard as important."

When Michael was sixteen, Jim fell in love with a twenty-one-year-old woman named Denise Garayalde, and they got married. It didn't sit well with Michael that his stepmother was only five years older than he was. One day he got into a quarrel with her. As he was walking past her, he flicked her hair aggressively with his fingers, in a gesture of rage. Jim responded by tackling his son to the ground, and he threw him out of the house, telling him that it was time for him to go and live with his mother. Eventually, that trouble blew over, and Michael went back to live with his father and Denise. He applied to Humboldt State University, in Arcata, and was admitted.

JIM AND MICHAEL DECIDED TO DRIVE TO ARCATA TOGETHER TO GET Michael installed at Humboldt State. Michael was eighteen. They drove in Michael's car, a hand-me-down diesel Volkswagen Rabbit that had once belonged to his grandfather. (Jim wasn't one to give his son a Porsche for graduation.) They were heading north on U.S. Highway 101, and Jim was feeling great about things, and also a little nostalgic: his oldest son was growing up, taking his next big step in life. But for now they were having a special time together, like the happy days in the family jet.

Suddenly Michael turned to his father. "Hey, Dad, I think I'd like to change my name."

"Change your name, Michael? Why do you want to do that?"

"I want to be thought of as a different person than Michael Taylor. What do you think?"

"I think it stinks," Jim said. He wondered if Michael was angry with him for some reason. "Look, Michael, you can have any first name you want. But changing your last name would offend me."

Michael said something to the effect that he wasn't angry or anything, he was just curious to see what it would be like to have a name other than Michael Taylor.

A few months later, Jim mailed Michael a check to help him with his college expenses. When the check cleared, he saw that it had been endorsed twice. Michael had signed it "Michael Taylor," and then he had endorsed it to someone else: "Pay to the order of Michael Mondragon." And he had countersigned it "Michael Mondragon."

Michael had changed his name! He had done it legally! Otherwise the bank wouldn't have let him sign the check that way. This stung Jim, who felt that his son had rejected him, and he thought the name Mondragon had a dark atmosphere around it, like something out of one of Michael's video games.

He telephoned his son. "Michael, what's going on? I told you it would hurt my feelings if you changed your name. Now I'll tell you what I think. I think you should find someone named Mr. Mondragon to pay your college bills."

Michael didn't say much in reply, but he went back to endorsing his father's checks with the name Taylor.

After he had been at Humboldt State University for two years, he decided to major in forestry, but he didn't sign up for many of the courses required for graduation, and Jim became concerned about this. At the beginning of each semester, Michael would sign up for required courses and everything seemed fine. But then he dropped those courses and signed up for courses that he really wanted to take, such as "Space, Time, and Relativity," and "Machine Language Programming."

"I'm not going to support you in college if you're not making regular progress," Jim said to Michael one day. "I expect you to graduate in four or five years."

Michael became interested in speed chess—chess that is played rapidly against a ticking clock. The game requires a keen sense of timing, an ability to sort rapidly through future moves, and the skill to aggressively exploit a hidden opportunity. He found that he had a real talent for it. He stashed a chess set and a chess clock in his Rabbit, so that he could play the game anywhere, with anyone. He started skipping lectures, and he got his friends to take notes for him.

With his friends attending college for him, he drove around the North Coast in his diesel Rabbit for days on end, and he went on long backpacking trips by himself. He avoided hiking trails and plunged straight into the wilderness. He would set up a campsite on some trackless spot by the Trinity River and fish for salmon or steelhead trout. When he was camping with his Rabbit, he would take out the chess clock and the chessboard, and he would challenge people at other campsites to play with him. Sometimes there was money riding on the game.

Jim Taylor was getting exasperated. One day in the fall of 1986, Michael called home to tell his father that he had lost interest in forestry and had decided to switch his major to engineering. Jim almost lost it. Michael had been attending college for several years, and now he was going to start with engineering? Trying to keep his voice calm, he told Michael that he wanted to see a list of *exactly* what engineering courses he had signed up for, and that he wanted to see this list *in writing*. "We're down to the nitty-gritty here," he said to his son. "If you don't take the courses you've promised you're going to take, then the die is cast and you're done."

On course-registration day at Humboldt State, Michael decided that he would spend the morning looking at some coast redwood trees that he was fond of. He would register for his courses in the afternoon, he thought. He lost track of time among the trees, though, and he ended up in Humboldt Redwoods State Park, where he spent a long time staring at a vast redwood called the Dyerville Giant, while the afternoon came and went. By the time he got back to campus, most of the engineering courses had filled up and were closed to enrollment. He did manage to sign up for two courses, and one of them actually was an engineering course. He mailed the list of two courses to his father. Mistake.

A few days later, Michael's telephone rang.

"Look, Michael, you have to take more units than this," Jim said.

"Well, I was sort of late registering for courses this semester," he replied. He didn't mention the Dyerville Giant. "But I am going to sign up for a bunch more engineering courses next fall, just before classes start."

"Well, you know what, Michael? I'm going to send you one more tuition check, for this fall semester, and that's going to be it." Jim told Michael that the following semester he would be on his own. He would have to start supporting himself and working his way through college. It was going to be a total financial cutoff. "I'm not giving you another dime after the end of this school year," Jim said to Michael. "When you go to graduate school, I'll pay for that."

It came as a shock, and as it sank in Michael burst into tears. This took place in front of his roommate, which greatly embarrassed him. Jim Taylor was just thirty-nine years old at the time, and Michael had just turned twenty. The incident left the two men deeply estranged. They were barely on speaking terms.

A FEW MONTHS LATER, ON THAT SUNNY DAY IN OCTOBER, WHILE HE walked up and down Redwood Creek looking at the Tall Tree, Michael Taylor was miserable, feeling that his father had cut him off in every way. Jobs on the North Coast were hard to come by, but he managed to land one making pizzas in a pizza parlor. This wouldn't pay for college, though, and he didn't know what he was going to do in life or even in a month or two, though it was very clear that he wouldn't be attending college anymore.

The Tall Tree stood about two hundred yards away, across the creek. Only its top was visible, poking out of the canopy. The Tall Tree—the Libbey Tree, the Mount Everest of All Living Things—looked kind of scrawny, Michael thought. It didn't look quite as magnificent as it was supposed to. As he squinted at it, he could see that the top of the tree had died and had snapped off. What was left of the top of the world's tallest tree had died back, too, and looked like a piece of broken driftwood. It was pretty apparent that the Tall Tree was no longer the world's tallest. Michael had been walking around lots of redwood groves, and he knew that he'd seen taller trees.

The tour group and the ranger were standing by the sign at the base of the tree, but they weren't actually looking at the tree—they were looking at the base of the tree and the sign there. In fact, from the place where they were standing, *they couldn't see the Tall Tree.* Most of it was invisible, hidden in the canopy.

It is extremely difficult to tell how tall a redwood is by looking at it from the ground. Redwoods that are closer can look very tall, when in fact they may be much shorter than the ones behind them, farther away. Michael Taylor had been staring at redwoods for years, and he had developed almost a sixth sense for knowing which ones were the tallest. He also had a sense for the size of a redwood; he could glance at a grove and see the true giants lurking in it.

Some of the trees in the Tall Trees Grove seemed to him to be taller than the Tall Tree. Many were much larger: muscular old nameless monsters with no signs at their bases. He had been around the redwoods long enough to know that in places the redwood terrain was nearly impossible to get through—a creased landscape choked with temperate jungle, where there were no roads or trails, no easy way in. He sensed the existence of blank spots on the map of North America—along the coast of California, no less. He had a strong feeling that the most inaccessible parts of the redwood forest along the North Coast had never been thoroughly explored. The world's tallest living thing was out there, somewhere, perhaps hidden in a lost valley. He could feel it. . . . He could just feel it out there, somewhere. He wondered if he could find it.

RUMORS OF A LOST CONTINENT

Humans are the only primates that do not spend time in trees. All other primates are arboreal or partially arboreal creatures. They live and move in forest canopies. Gorillas are the largest primates, and they live mostly on the ground, but even they spend significant amounts of time in trees—a gorilla can spend up to 20 percent of its day in trees. Gorillas have prehensile feet; they can grip a branch by closing their foot around it, using an opposable big toe. Human children climb trees naturally and rather easily, especially when they are barefoot, but as children grow older, as their bodies grow larger, taller, and heavier, and their feet become stiffer, they seem to lose their natural affinity for tree climbing.

The earliest primates appeared in middle Paleocene times, around sixty million years ago. This was early in the age of mammals and not long after the disappearance of the dinosaurs. The earliest primates were squirrel-like creatures whose hands and feet had opposable thumbs and toes, making it possible for them to grip branches. They had long tails that were whiplike and flexible in the beginning, and gradually evolved to become prehensile and capable of gripping branches. Roughly six million years ago, a population of apes (tailless primates) living in the forests of the Rift Valley region of Africa began to evolve away from the trees. These were the ancestors of the human species. They began foraging on the ground, moving through the grassland savannas that were developing in East Africa. They evolved into hominid apes that walked upright, with knees and feet that were suited for traveling long distances but were not good for climbing trees. The use of tools as weapons probably helped these hominids survive on the ground.

At some point, the ancestors of humans must have developed a natural fear of heights—a fear of being up high in a tree. Humans are the only primates I know of that have an inborn fear of heights. Other primates, when they are frightened, instinctively run up a tree, where they feel safe and at home. Hominids who felt insecure in trees, more afraid of heights, and perhaps more willing (in their behavior) to move out across open ground may have had a better chance of surviving and producing offspring. Open ground would have seemed as terrifying to many primates as heights do to many humans. In any case, the forest canopy from which we had arisen became as remote from us as the depths of the sea.

IN THE EARLY 1970S, WILLIAM C. DENISON, A NATURALIST AT OREGON State University in Corvallis, began climbing up into tall Douglas-firs to see what was going on there. Denison used a rope and Jumar ascenders when he climbed. Jumar ascenders are devices with handles that a climber can hold, and teeth that grip a rope. Jumars are used in pairs for climbing straight up a rope. The Jumars are attached to a harness, which a climber wears around his hips and thighs, and which prevents the climber from becoming detached from the rope and falling. The rope-climbing method using a pair of Jumars or similar mechanical rope ascenders is a kind of one-two motion that climbers call jugging. Mountaineers and cavers use Jumars for climbing up a fixed rope that's hanging vertically down a cliff or along the wall of a chasm in a cave.

Bill Denison wore a primitive sort of belt around his waist—not really a climbing harness at all—and he was able to jug up a couple of hundred feet into a tall Pacific conifer in a half hour or so. Once he had gotten into the tree's top, he collected fungi as well as lichens from the branches. Denison hammered bolts into the trees as he climbed them. He bolted his way up tall trees in Oregon, sometimes climbing alone, and sometimes with a colleague, Lawrence Pike. He enlisted the help of physically fit college students to assist him in collecting samples, as well as his son, Tom Denison. (It seems almost a miracle that none of these people got killed, given how sketchy the equipment was.) In 1973, Denison wrote a famous account of his tree-climbing adventures, "Life in Tall Trees," in *Scientific American*. "A treetop in a forest, like a mountain peak or a deep canyon, is a remote world that is plainly visible but not easy to explore," he said.

One day around the time Bill Denison was jugging up trees in Oregon, a French botanist named Francis Hallé took a walk in the rain forest in French Guiana with a group of students, and they stopped to rest, sitting on a rotten log. As Hallé recalled in an interview years later: "We were looking at the tree canopy—so many epiphytes, so many animals. Remember, it was the period of the Apollo flights. A student said, 'Funny! Man is able to collect stones on the moon but unable to work in the canopy.' We started discussing this. 'How can we do that?' 'Helicopter?' 'No, it's too noisy, too expensive, too dangerous.' 'So, how?' 'Dirigible?' General hilarity, then silence. The idea was born."

Francis Hallé and two colleagues, Dany Cleyet-Marrel (an aeronautical engineer) and Gilles Ebersolt (an inventor), came up with a dirigible flying machine that consisted of a hot-air blimp under which dangled a huge inflatable raft. The raft was called the *Radeau des Cimes* (the Canopy Raft). The French scientists would launch their dirigible and rise above the rain forest, riding in the *Radeau des Cimes*. The airship's engines weren't very powerful. They would putter along above the rain forest, looking down into its secret spaces, and when they saw a tree of interest they descended until the raft was lying on top of the canopy. Now they were floating upon a green ocean in a raft.

The bottom of the raft was made of netting. The passengers could reach through the netting and gather samples, and look around in the canopy. When they were done, the dirigible would lift up the *Radeau des Cimes* and go on its way. The *Radeau des Cimes* usually sat on top of one tree for a week or two at a time. Hallé wanted to move around the canopy more quickly, so he and the engineers invented a second dirigible flying machine they called the Sledge. It was more maneuverable, and it floated from tree to tree. Francis Hallé's flying machines were brilliant inventions, and they had a distinctively French air of *fantaisie,* or clever whimsy. (The hot-air balloon was invented in France by the brothers Joseph and Jacques Montgolfier. In 1783, the Montgolfier brothers launched a balloon in front of King Louis XVI and Marie Antoinette. It carried a rooster, a sheep, and a duck for two miles and landed the animals safely—rather fantastic.) The French method of canopy exploration didn't require anyone to know anything about climbing trees, although blimp-handling skills were essential. Francis Hallé eventually became semi-retired, and the dirigible voyages to study the canopy were somewhat curtailed for lack of funds—a common problem in canopy research.

In 1978, Margaret D. Lowman, a young American graduate student in botany at the University of Sydney, in Australia, decided to write her dissertation on treetops. She had been anxious about choosing a topic, and she thought that at least nobody had tried this one. Lowman wanted to climb the trees, but she had no idea how to do that. She joined a caving club in Sydney, and the other members taught her how to climb a rope using Jumar ascenders. Lowman sewed a climbing harness for herself made out of seat-belt straps, and welded some pieces of iron together to make a slingshot. She then went into a forest near Sydney and used the slingshot to shoot a fishing line over the branch of a tree, after which she attached a thin nylon cord to the fishing line and dragged the cord over the branch. Then she attached a rope to the nylon cord and pulled it over. Lowman began making solo ascents into the rain-forest canopy of eastern Australia. "When I first started out climbing trees, I had no idea that they held fifty percent of the life on the planet," Lowman said to me. "We had no clue that the forest canopy is this amazing hot spot for biodiversity."

Meg Lowman began studying insects in the rain-forest canopy. She also began studying leaves. She wanted to know how long the leaf of an Australian rain-forest tree might last. Many tropical and subtropical trees (and some temperate-forest trees, such as live oaks and magnolias) do not lose all their leaves annually, as deciduous trees do. Lowman used a Magic Marker to write numbers on the leaves of some Australian trees, and then she climbed up into the trees every so often to see how many numbered leaves were still hanging there. "I'm from upstate New York, and I figured maybe six months, and then the leaf would fall off," she said. Nineteen years later, entering middle age, Lowman found leaves with Magic Marker numbers on them that she had written on the leaves as a younger woman. The leaves had remained alive and unchanged for almost two decades. This illustrates the difficulty humans can have in seeing what's happening in a forest canopy. Humans don't live long enough to see many events in trees unfold. Lowman had spent much of her career trying to observe the fall of a leaf.

While she was still working on her dissertation, Meg Lowman was standing on the branch of a tree in New South Wales one evening. It was getting dark and a thunderstorm was brewing, and she wanted to get down before the lightning came. In her haste, she removed the Jumars from her rope before attaching a descender device to it. This meant that she wasn't attached to the rope: she wasn't attached to anything—she had lost her physical connection to the tree. She slipped and fell off the branch, and went into free fall. She fell fifteen feet, hit the ground, and found herself lying stunned, badly bruised, but with no broken bones. What am I doing? she asked herself. She had been dating a local sheep rancher named Andrew Burgess. Lying there on the ground, Lowman decided that she had better get married and settle down.

Lowman and Burgess had two sons. Eventually, she found that being the mistress of an Australian sheep ranch was incompatible with exploring the canopy, and she left her husband and returned to the United States with her children. She ended up as a professor at the New College of Florida, where she began to focus her efforts on designing and constructing canopy-access structures to help her students get into the canopy safely and easily—towers, suspension bridges, stairs. Not long ago, the astronomer Carolyn Shoemaker, who is the world's leading discoverer of comets and asteroids, named an asteroid after Lowman. ("I love to name real estate in outer space after women whose work I admire," Shoemaker said in a letter to Lowman.) Minor Planet Lowman is a lump of stone the size of Mont Blanc, and it follows a peculiar orbit near Jupiter.

What William Denison and Margaret Lowman were doing to gain access to the forest canopy is called direct climbing—you climb the tree using ropes and your body. In the late seventies, while Lowman was learning how to get into trees in Australia, a young woman named Nalini M. Nadkarni, a graduate student in ecology at the University of Washington, was in Costa Rica studying a tropical forest there. As she looked around at the trees, she realized that she couldn't see much of the forest, and it occurred to her that she needed to climb into the trees in order to study them properly. A biologist named Donald Perry taught her the basics of direct climbing with a rope and Jumars, but she met with some skepticism from other biologists. As Nadkarni explained to me, "People said, 'What do you mean, you're going up into the trees? There's nothing up there. That's just Tarzan and Jane stuff.'" She believed that she was onto something. "I felt like, 'Wow, here's this new place nobody's been to. We might find new species up here, new interactions in nature.'" Nalini Nadkarni became a professor of biology at Evergreen State College, in Olympia, Washington, and an internationally famous forest-canopy scientist.

Climbing is not the only way to see the canopy. Another way is to cut down a tree and study everything in it while it's lying on the ground—it's a sort of autopsy of the canopy, and it provides a detailed look at a tree, but it doesn't reveal processes of life that unfold in the tree over time. Another thing researchers do is put an insect-fogging machine under a tree, which sends up a cloud of smoke that knocks out the insects in the tree, and they fall to the ground and can be studied and counted. Researchers can also simply look at the canopy through a pair of high-powered binoculars, or they can send up balloons carrying a camera or data-gathering instruments. These balloons are like small space shots into the trees. A forest canopy can also be observed from aircraft and from satellites in orbit.

AT FIRST, BIOLOGISTS FOCUSED MUCH OF THEIR ATTENTION ON TROP-ical rain-forest canopies. Tropical forests all over the earth are being destroyed by logging, farming, human-made fires, and generally by the great mushrooming of the human population that has taken place during the past hundred years. In 1900, the world's population was about one and a half billion. By 2000, it had risen to more than six billion and was heading for ten billion. The human population has pushed into the tropical rain forests, and they are disappearing.

As people enter tropical forests, they come into contact with species of animals that harbor infectious organisms, viruses that can move into the human species, as HIV has done (coming from chimpanzees or other primates that live in the rain forest) and as Ebola virus seems to be doing (coming possibly from bats that live in caves or in the forest canopy, though no one knows). The tropical rainforest canopies house an incredible variety of living things, and biologists wanted to try to identify what lived there and to get some idea of how the ecosystems worked before the forests ceased to exist. They also felt that they could help save the rain forests if they could convince people that the forests were beautiful and abundant with biodiversity, and therefore were worth preserving, both for the good of the human spirit and for the economic and physical well-being of civilization in future times.

Temperate rain forests, which occur in cool but not cold climates, and which include the tall rain forests of the Pacific Northwest, received less attention. Temperate rain forests typically consist of conifers—tall, slender evergreen trees. The temperate rain-forest canopies were more difficult to climb into, and they were thought to be less interesting, with less variety of life. Tall temperate forests are far taller than tall tropical forests. The tops of trees in the Amazon rain forest are typically between 120 and 150 feet above the ground. These trees are very tall, but they are only half the height of a typical old-growth Douglas-fir or Sitka spruce rain forest in the Pacific Northwest. Tropical rain forests are also less massive than temperate rain forests. Biomass is the sheer weight of living material in an ecosystem. The biomass of a temperate rain forest on the Olympic Peninsula of Washington seems to be two or three times greater than that of the deep rain forests of the Amazon Basin.

Clearly there was a lot of life in the temperate rain forests, but biologists hadn't really focused on these forests. They assumed that a temperate rain forest's biomass was mostly just big trees. In any case, the virgin temperate rain forests in the continental United States had been almost totally wiped out by logging, and biologists may have concluded that there was very little left to look at. The rain forests of Canada and southern Alaska were also essentially being stripped away by logging, as were the temperate forests of Chile and southern Australia.

The remaining redwood forests were thought to be virtually unclimbable. The redwoods were visibly dangerous—scary in the extreme, intimidating as trees. A redwood typically does not have any strong branches on the lower part of its trunk, which is just a huge, bare column, sometimes feathered with little branches, stretching upward to an impossibly high canopy. The lowest strong branch on a redwood may be 250 feet above the ground—twenty-five stories up. Well into the last decade of the twentieth century, the redwood forest canopy of California was one of the last unseen realms of nature on the planet. Nobody had entered the redwood zone above the level of the ground, except for a college kid named Stephen C. Sillett, who had nearly gotten himself killed there.

THE NAMING

IN THE FALL OF 1987, WITHIN A WEEK OR SO OF WHEN MICHAEL Taylor realized that the redwood forest had never been fully explored on the ground, Steve Sillett realized that it was unexplored in the air. It seems almost incredible that two college juniors who didn't know each other simultaneously concluded that parts of California remained off the charts. The earth's tallest forest was a shrouded mystery.

As soon as the fall break was over and classes had resumed, Sillett walked into Professor Bert G. Brehm's office in the biology building on the Reed College campus. He placed a plastic bag full of bits of moss and lichen on the professor's desk. Bert Brehm was an extremely popular teacher at Reed and a distinguished scientist. He taught botany, and he had a laboratory, cluttered with gizmos and gadgets, where he did experiments on plants. The floor of his office was piled with books and papers. With white hair and a white goatee, Brehm had an elegant appearance, and he spoke in a mellifluous voice. Sillett was awed by him.

Sillett picked his way around the piles of paper on the floor and opened the bag, spilling the bits onto Brehm's desk. "Dr. Brehm, what are these things?" he said.

"They're lichens, obviously, and bryophytes"—mosses—Brehm replied. "Where did you collect them?"

"At the top of a redwood tree."

"What's this all about?"

"Dr. Brehm, I made a flying leap into a giant redwood, and I climbed the tree to the top. I found out that I can actually jump from tree to tree. I don't think anyone's ever done that before."

Brehm learned that Sillett had performed this remarkable feat with Marwood Harris, and Harris wasn't even a biology student. "Hold on, Steve," he said. "I want to say that this sounds appalling. What you did was very dangerous."

"But, Dr. Brehm, I'm capable of judging my own safety. I wouldn't take any unnecessary risks."

This was disturbing to Brehm. What, exactly, did his student mean by "unnecessary" risks? Did Steve mean to say that jumping around in redwoods was a *necessary* risk? "You led another Reed student up that tree with you," Brehm said. "Someone could die climbing trees this way, and it might not be you." Brehm didn't think that the reality of death had registered with Steve, but he suspected there wasn't much he could do about it just then. The lichens, however, were quite interesting. He took down a handbook on lichens, and he and Sillett pored over it together, trying to determine their names.

One of them turned out to be *Cetraria orbata*. This is a kind of Icelandic lichen that's common in California, though nobody had ever collected it from the top of a redwood. The lichen is grayish green in color, with a frilly shape and hairlike structures growing around its edges. Another one was called *Platismatia glauca;* its common name is ragbag. Ragbag grows on the branches of trees, where it looks like a wad of old rags stuck to the branch. You can find it in trees in Scotland, as well as in California. It doesn't occur on the ground. There was another one called bone lichen—it looks like bleached bones. There was a beard lichen and a witch's hair lichen. None of the lichens were unknown, but their occurrence together in redwood trees was something previously unknown. They were what biologists call an assemblage; a community of living things, existing in a habitat, a place in nature all its own.

Bert Brehm was a friend of Bill Denison, the naturalist at Oregon State University who had started climbing tall Douglas-firs fifteen years earlier, gathering fungi and lichens. Brehm decided that Steve Sillett needed to meet Denison so that Denison could give him a lesson in the safe way to get up a tree. Sillett didn't have a car and didn't know how to drive, so Brehm drove him two hours to Corvallis to meet the famous tree climber.

Bill Denison invited the younger man to climb a Douglas-fir. A week or so later, he drove his truck to Reed College, met Sillett at his dorm, and they went into the Cascade Mountains. They became friends during the drive. Denison took Sillett to a very tall Douglas-fir that grew in the H. J. Andrews Experimental Forest, on the Middle Santiam River. The fir had a rope strung up in it. The rope had been hanging in the tree for ten years, baking in the sun, and it was covered with mildew and moss, and was frayed, weak-looking, and sketchy in the extreme. While Denison watched from the ground (he was getting old), Sillett put on his rock-climbing harness and climbed up the tree along the sketchy rope. Bill Denison, unbeknownst to Bert Brehm, had a blithe attitude toward matters of safety.

Denison had become fascinated with a species of lichen called *Lobaria oregana,* or lettuce lungwort, which grows in the tops of old Douglas-firs. It is bluish green, and has a shape like curly lettuce. Lobaria lichen consists of a fungus combined with a type of cyanobacterium. Cyanobacteria collect sunlight for energy (as do plants), and they have the ability to take nitrogen directly out of the air and incorporate it into themselves. (Plants are unable to do this. Fungi called mycorrhizae are attached to their root systems, and these fungi take nitrogen out of the soil and supply it to the plant. It's another partnership.) Lettuce lungwort pulls nitrogen straight from the air and incorporates the nitrogen into the lichen.

Nitrogen is fertilizer.

Bill Denison had discovered that lettuce lungwort provides old Douglas-fir forests with large amounts of fertilizer. The lungwort doesn't occur in young trees. Where it does occur, in old forests, it fertilizes the forest spontaneously, producing fertilizer from the air—fertilizer for free, from nothing. When a piece of lungwort falls off a branch (lungwort is constantly torn from branches in storms), it falls to the ground, where it can't live, and it dies and rots away, dumping fertilizer into the soil. Lungwort can also die while it's stuck to a branch, and it rots in place, providing fertilizer to epiphytes that occur on the branch itself, feeding mosses and ferns and other plants that live in the air. The additional fertilizer in the soil and in the canopy helps all the plants and trees grow larger, and they, in turn, are able to support a greater variety of insects, birds, and mammals. Denison was seeing the underpinnings of an entire forest ecosystem in his studies of a humble little thing that looks like lettuce and never grows on the ground.

Denison, however, had gotten distracted. He loved wild mushrooms, and he had supported himself in college by gathering them and selling them to restaurants. He was especially fond of the shiitake mushroom, which grew in Asia but wasn't found in the United States. Recognizing that shiitakes might be worth a lot of money in this country, he studied Asian mushroom-growing practices and developed a way to farm shiitakes and other wild, rare mushrooms commercially. He founded a company called Northwest Mycological Consultants, and he basically created the market in the United States for exotic farmed mushrooms. Denison became so fascinated with the problem of how to grow rare mushrooms and sell them that he didn't pursue his research on Lobaria to very detailed conclusions, or publish all of his findings in top journals. He ended up with a sort of underground reputation among forest ecologists. As for his work with mushrooms, Bill Denison is the main reason that people are able to buy shiitakes in a supermarket.

Denison drove Sillett back to Reed College after he took him out climbing that first day, and during the drive, as Sillett listened to the older man talk about what Lobaria, or lungwort, does in a forest, he decided that he wanted to spend the rest of his life studying Lobaria in the trees.

WHITE PINES

MARIE ANTOINE'S FATHER, RONALD, DIDN'T KNOW HOW much time Elizabeth, Marie's mother, had left. When he spoke with Marie about her mother's bone cancer and what the future might hold, he never spoke in terms of time. He avoided the subject of time as much as possible.

It became apparent that Elizabeth couldn't live on Treaty Island during the winter. She was losing the ability to climb stairs, and the path that went from the boathouse up to the Cottage was steep, and it got icy. The island was cold and windy, and too far from town and from medical help. The best way to get to Treaty Island in the winter was to drive a car across the ice, but that was a little tricky, and the family didn't own a car. (Ronald mostly used an aluminum outboard motorboat in place of a car.) Whenever Ronald wanted to cross the ice, he borrowed a car from one of his friends and drove across; the ice could be as much as six feet thick in the winter. Eventually, he bought a town house for the family in Kenora, the largest town on the shore of Lake of the Woods, a few miles away from

Treaty Island, and the family took up residence there during the winter months.

On school mornings in the fall, before the weather got too cold and the family moved into Kenora, Marie would go down to the boathouse with her father and get into the family's aluminum boat, and he would take her to school. When school let out in the afternoon, her father would be waiting for her at a landing in the town, and they would drive back home. The old Johnson outboard motor would sputter and hum as they crossed glassy water on those dark afternoons. The pines were outlined against the bluffs. Below and among the pines you could see splashes of red maples, like torches burning in the last gray light of afternoon, and the lake had a musty smell.

They would climb up the bluffs to the house, where Elizabeth sat in her chair by the living-room window. She didn't want Marie and her little sister, Bella, to see her lying in bed, so she usually got into the chair in the afternoon. Ronald helped her out of bed and put soft blankets around her to keep her warm. He gave the children a snack, and afterward he read aloud from a book, while the girls lay curled up on the rug at their mother's feet. They loved it when he read to them from *Watership Down*.

Rabbits . . . are like human beings in many ways. One of these is certainly their staunch ability to withstand disaster and to let the stream of their life carry them along, past reaches of terror and loss.

Afterward, Ronald would cook dinner for everybody, using vegetables from his organic garden. Marie would practice the piano after dinner, and read, and she and Bella would play with each other in their rooms, making houses out of chairs and furniture and having imaginary activities with their dolls. There was no television in the house. At the end of the evening, Ronald would put Elizabeth and the children to bed, kissing them all good night. He was usually the last person awake in the house, except when Elizabeth couldn't sleep because of the pain.

ELIZABETH HAD TO GO TO THE HOSPITAL IN KENORA ONCE IN A WHILE, when she was having one of her downturns. When her mother was in the hospital, Marie would get on a city bus after school and take it across town to visit her. She would find Elizabeth propped up in bed, smiling and laughing. Nurses always seemed to be standing around and listening to her stories. Elizabeth had a ribald sense of humor, and she loved dirty jokes, but when Marie came into the room she changed the topic. She didn't seem to be terribly discouraged by her condition, even as the bones inside her body became twisted and developed holes in them.

Marie ate snacks in the hospital room. Ronald didn't like to throw anything away. He gave her bags with snacks in them, and milk in plastic containers that had disposable plastic straws. He washed the straws so that they could be used again, but he could never get them clean, and they smelled like sour milk. He explained to Marie that he had proposed to her mother on the island, and that they had had nine beautiful years together before she got sick. They were still having beautiful years together, he said.

Once a month or so, Elizabeth had to have treatments at a large hospital in Winnipeg, which is a three-hour drive from Kenora. Ronald could have borrowed a car from someone, but he wanted to be able to look after Elizabeth during the trip, so the family would take a Greyhound bus to Winnipeg, and take another one home after Elizabeth's treatment. During the long rides, Marie did her homework or looked out at the Trans-Canada Highway going past. By the time she was six or so, she had become a very self-contained person.

"I remember developing a certain feeling when I was young," Marie Antoine said to me one day. "It was a feeling that I had to be good, and that I wanted to be good. I wanted to be well behaved and get good grades in school, because it would make things easier for my parents. Even when I was five or six, I could see how much they loved each other, and I felt so sad for what they were going through."

Ronald finally bought a television for the town house, because the children needed something to do in the winter evenings with Elizabeth, and the television didn't demand much of her. Marie liked to lie on an afghan on the floor while she watched television at her mother's feet—it was one of her mother's wrap-up blankets, and it was especially soft. "One night in our town house, when I was six or seven years old, I was watching television with my mother, and I had this sudden urge to throw up—you know how it is with little kids," Marie said. "I projectile-vomited all over the afghan. My poor mother, as sick as she was, ended up trying to clean up the afghan, because she didn't want my father to have to do it. I helped her. We put it in the dryer, and we were amazed when it came out a quarter of its original size. I still have the strangely shrunken afghan. It's a memory of my mother."

Marie's father seemed happiest when he was outdoors. He would take clippers and a saw and clear the paths, and remove brush from around the white pine groves on their property. He often asked Marie to help him with the work, and she would clip and trim branches, and carry or drag bunches of sticks to piles. She hated the bugs and the heat, and was quite bored by it all, and she pretended to be working when she wasn't, or sometimes she slipped away and went for a swim. Ronald didn't get mad, and he just kept working, steadily. Elizabeth was never well enough to walk on the paths Ronald cleared, but almost every day, if she could, she would go onto the deck of the Cottage, which was under the white pines, and look out from the cliff over the lake. The two of them would sit by themselves and talk. The view encompassed dozens of islands going off into the horizon.

In the spring of the year that Marie was eight, they weren't sure that Elizabeth was strong enough to visit the island at all, but Ronald had a flight of wooden steps built running from the boathouse landing up to the Cottage, with benches placed every twenty-five steps, so that Elizabeth could rest as she climbed. In early May of that year, he finally told Marie that her mother might die. He had never said this to her before. It was something that Marie hadn't thought of. She knew that her mother was sick, but the idea that her mother might stop being there came as a surprise to her. Elizabeth's doctors, however, had told her that the cancer treatments had failed, and it was only a matter of time now. The best anyone could do was to help her be comfortable.

They did get to the island, and when Elizabeth could no longer walk Ronald pushed her in a wheelchair. Marie and little Bella would get restless waiting for their father to push their mother up the wooden steps, and to pass the time they took to climbing a white pine tree that grew out over a ramp that led into the Cottage. They would drop pinecones down on Elizabeth as Ronald pushed her. Marie moved easily up and down the tree. Bombing their mother with pinecones, she and Bella made her laugh. "Cut that out, you two!" Elizabeth would say, but she loved it. Elizabeth Antoine died on June 4, 1984, and was cremated. Ronald, Marie, and Bella scattered her ashes on a rocky outcrop in a stand of white pines above the Cottage, with a beautiful view.

MARIE AND BELLA ARE LYING ON THEIR BACKS ON THE ICE NEAR Treaty Island. It is a moonless winter night. Twenty degrees below zero. No wind, not a breath of moving air. They are visiting the Cottage; Ronald Antoine has borrowed a friend's car and driven it out there on the ice, and he is up at the Cottage doing some work. The children are lying on the ice, waiting for their father to finish whatever he's doing.

Lake of the Woods is asleep for the winter, but it is dreaming. Marie feels that she can hear the dreams of the lake running through the ice, like thoughts in a language we don't know. The ice begins to creak. It makes banging noises, and groans, and makes little pings and snippy-snap sounds, and sometimes there are long, drawn-out booms, like cannonfire heard from a distance. The cracks and adjustments in the ice can be heard racing from island to island and across the bays, traveling for miles and moving very fast, giving out stereo sounds. Apart from that, the world seems quiet, without wind, without any clicking or rustling of branches, without any sound of a living thing. Overhead, the misty river of the Milky Way turns slowly with the handle of the Little Dipper around the North Star. The North Star is motionless, and everything else in the sky is moving.

The night sky shimmers with greens and whites and pinks—the northern lights. Sometimes the northern lights billow up, and they flare and shine across the sky from north to south, from horizon to horizon. The northern lights happen in complete silence. It is as if some wonderful celebration is occurring in a faraway place where you can't hear it or see what's really happening, but you can be happy just knowing about it. A meteor crosses the sky—just a *zip* in the corner of Marie's eye, and it's gone.

She had always wondered about the sky. What really is the sky? How far does it go? she was thinking. What's beyond it?

The night was growing colder, and the ice was making ringing sounds, like church bells pealing. And then, mysteriously, frighteningly, all the sounds of the ice stopped, and there was complete silence. Marie felt almost overwhelmed by the stillness within the lake, a silence so profound that she could feel it inside her body. It was as if time had stopped. This never went on for long. After a minute or two, the lake started making noises, and time began moving again.

THE GROCERY CLERK

AT CHRISTMAS 1987, TWO MONTHS AFTER HE VISITED THE Tall Trees Grove and realized that the world's tallest tree had never been found, Michael Taylor dropped out of Humboldt State University. He put his worldly belongings into his diesel Rabbit and drove it south to San Diego, where he got a job with the Cutco Cutlery Corporation. Taylor signed on as a door-to-door knife salesman, and he got some training to get started. He requested Santa Barbara as his territory and rented a room in San Diego; he planned to drive up to Santa Barbara on the weekends to sell knives. As a marketing gimmick, he got some black and white paint and a roller brush, and he painted his Rabbit with zebra stripes. Then he headed for Santa Barbara with a suitcase full of Cutco knives.

The Zebramobile's first stop was Hope Ranch, where he parked in the driveway of his old home. He asked his father and stepmother if they'd like to see him demonstrate his sales pitch; he felt that he needed to practice it.

Jim Taylor was a little skeptical. He hadn't spoken with Michael that often recently, and he wasn't quite sure what Michael was up to.

He and Denise sat in the living room and watched while Michael opened his salesman's suitcase and spoke of the excellence of Cutco knives. He was quite persuasive. Somewhat to his surprise, Jim Taylor found himself writing a check to his son for five hundred dollars' worth of cutlery.

- "Dad, I was wondering if you have any friends who might be interested in these knives," Michael said, after he had closed the sale and pocketed the check.

Jim was delighted with Michael's performance. He went into his study and got out the members' directory to the La Cumbre Country Club, in Santa Barbara. This is one of the most exclusive country clubs in the United States. The members' list is available only to members of the club, and it's full of the names of wealthy and famous people. Michael had caddied at the country club when he was in high school, and he knew quite a few of the members, and Jim reasoned that it would be all right to share the list with him, quietly.

Using the names as a hot list, Michael began making sales calls on the members of the La Cumbre Country Club. His Zebramobile became a known oddity around Santa Barbara's better neighborhoods. It could be seen circling through the mansion-studded hills of Hope Ranch, trailing diesel smoke and bottoming on its shocks. Michael's hair had gotten extremely long, and it hung down to his waist. He was often stopped and questioned by the police, but he found that he could talk his way out of this kind of situation by being friendly and explaining that he was just a Cutco sales representative. It also helped to casually mention one or two well-known names as being among his customers.

One Sunday afternoon, Michael parked his car in the driveway of Alan and Cindy Horn. Along with a few partners, Alan Horn had recently founded Castle Rock Entertainment, an independent film and television studio, which would soon go on to have a smash hit with the television series *Seinfeld*. Alan Horn would eventually become the chairman of Warner Bros. Studios and one of the most powerful figures in Hollywood. On that afternoon, Michael Taylor lifted his salesman's suitcase out of his striped car and rang the Horns' doorbell. In the living room, while the Horns watched with a certain polite interest, Michael opened his suitcase and pulled out a pair of cheap scissors. Not Cutco scissors. Next, he lifted out a hunk of gnarly rope as thick as a salami, along with a slab of dry leather that looked as if it had been ripped out of a mummified ox. "Consider these sample materials," he said. "Now let's see what happens when we use these typical scissors on them." He made a production of chawing and gnawing the cheap scissors on the rope and the leather, but the scissors wouldn't dent them. Then, from his suitcase, he lifted out a pair of Cutco Super Shears—gleaming überscissors. He closed them on the leather, and it fell apart like egg custard. "Nothing cuts like a Cutco edge," he said, and then, with one easy snip, he cut the salami rope in twain. But wait. He held up a penny and cut it in half with the scissors, as if it were made of Play-Doh. He handed the pieces of penny to the studio executive and his wife so they could inspect them. Next, out of his suitcase came apples, oranges, and tomatoes, and he cut them into bits while asking his prospects to note that these tomato slices had a thinness that rivaled paper. The Horns bought nearly a thousand dollars' worth of knives and scissors from Michael Taylor.

On another day, he eased the Zebramobile into the driveway of Fess Parker, who had become an icon playing Daniel Boone in a long-running television series, and also Davy Crockett in the Walt Disney movie *Davy Crockett: King of the Wild Frontier.* Parker had later made a fortune in real-estate development around Santa Barbara, and he founded the Fess Parker Winery and became one of the more respected winemakers in California. Parker remembered Michael Taylor from the golf course at the country club, and he welcomed him into his home and asked him what he was up to these days. Michael cut a penny in half in front of Fess Parker. He asked Parker to inspect the unique design of Cutco's exclusive Double-D® serrated edge. "The Double Diamond edge keeps blades sharper longer," he explained. "It is the most superior serrated cutting edge ever made."

Fess Parker evidently wanted to know how long one of these blades would stay sharp.

FOREVER®, Michael explained, because every Cutco blade came with the famous and unique Cutco Forever guarantee. Parker or his heirs could bring his Cutco knives in to any Cutco dealer for sharpening or replacement, anytime, free of charge. "Even if you return the knife with just a fraction of the blade left on it, they'll give you a new one, no questions asked," he said. As Daniel Boone, Fess Parker used a bowie knife for his cutlery needs, which included carving up bears and morally rotten white men. At home, however, he might possibly have preferred Michael Taylor's knives. (When I called Fess Parker to ask him about it, he said through a spokesperson that, regrettably, he just couldn't recall what sorts of knives he was using in his kitchen around 1988.)

Jim Taylor felt a little uneasy about sharing the members' list with his son for sales purposes: it was the kind of thing that could get you thrown out of the country club. He had been wondering if he would start hearing complaints about how Michael had been pressuring members to buy the knives. Instead, people thanked Jim, saying what a great salesman his son was and how happy they were with their new knives.

Michael Taylor sold a knife to his mother, Cassie Jane, who promptly cut herself with it and had to go to the emergency room to get stitches. He sold one to his brother, Jeb, who also had an accident with it. ("Those were incredible knives, but we were cutting the shit out of our fingers with them," Jeb told me.)

One day he showed up at the house of Roscoe Tanner, the tennis star. The doorbell was answered by Charlotte Tanner, Roscoe's wife. Roscoe wasn't at home, but she greeted him warmly. He said that he was sorry to miss Roscoe, but would Charlotte like to learn more about Cutco knives? She couldn't really say no. In the end, she purchased the Homemaker Plus Eight knife set, the Super Shears, a number of smaller Cutco scissors and shears, and a complete set of Cutco kitchen tools. Roscoe Tanner returned home to find that Jim Taylor's son had sold his wife between two and three thousand dollars' worth of cutlery.

"Roscoe Tanner practically had a heart attack," Michael Taylor said. "He was really pissed off, and for a while he wouldn't speak to me. Cutco does make a very good product, though. I could be proud of selling their stuff." Charlotte was happy with her new knives, and Roscoe eventually got over it and he and Michael have remained on friendly terms. (Roscoe and Charlotte Tanner divorced several years later. She kept the knives.)

In the space of six months, working mainly on weekends, Michael sold sixty thousand dollars' worth of knives to rich people around Santa Barbara, and was promoted to assistant district sales manager in the Cutco sales force.

Jim Taylor was proud of Michael's success, and told him so. Michael maintained his distance, though. Apart from the sales demonstration, he recalls speaking on the telephone with his father only once that year: he called Jim to wish him a happy birthday.

Even in the hands of a sales genius, the door-to-door knife business has its natural limits. Michael's cutlery deal-flow began to falter. He had saturated Santa Barbara's élite households with Cutco knives, and they were guaranteed forever. Eventually, the rich people around town didn't need any more knives or scissors. He realized that he had to broaden his customer base. He painted out the zebra stripes on his car, covering them with gray primer paint, so that he would look a little more respectable, and he began going downscale, driving into middle-class neighborhoods in search of new customers. He started ringing doorbells in tract developments, where the police didn't know him. The fact that he was selling knives out of a car covered with primer paint, and that his hair was longer than Charles Manson's, didn't go over well. ("Try knocking on strangers' doors and saying, 'Can I come in and show you a bunch of knives?' It doesn't work," Taylor explained to me.) His business went into a death spiral.

He abruptly went from being a top star in the sales force to a struggling has-been. His Rabbit became affected and began dropping small loose parts under the strain of road mileage, the fruitless circling in neighborhoods. "Every time I started my car, it seemed like another bolt fell out of it," Michael said. "I stopped trying to figure out where the bolts were coming from."

He decided to become a rare-coin dealer.

He quit his position at Cutco and started hanging around in coin shops, chatting up the dealers and getting familiar with rare coins. Then he began buying and selling silver coins—trading the coins with dealers, using the money he had saved from his knife business. He found that he had a talent for spotting an opportunity in the coin trade. He had a very sharp eye. One day he walked into a coin dealer's shop and bought an 1896-S Barber quarter for twelve dollars. It is one of the rarer American quarters. Apparently, the dealer hadn't examined the coin closely, and hadn't noticed a somewhat faint *S* on the back of the quarter below the eagle. The *S* is what makes the coin rare. A few days later, Michael sold the quarter to another dealer for seven hundred dollars.

Finally, he gave it all up, packed his Rabbit with his worldly possessions, including some silver coins, and he drove back to Northern California. He took out some college loans and reenrolled at Humboldt State, paying part of the first semester's tuition with his savings from selling knives and trading silver coins, and with a tiny bit of money from his winnings at speed chess. He still had to earn a living. He got a job as a grocery-store checkout clerk at the C&V Market in Eureka, a small city ten miles south of Arcata.

THE C&V MARKET WAS A MOM-AND-POP CORNER GROCERY STORE IN a dingy neighborhood. Michael Taylor's pay was four dollars and fifty cents an hour. ("I was getting twenty-five cents above minimum wage, and I was glad to get it," he said.) With a steady income at last, he was able to live modestly and continue with his college studies. Even though he was enrolled at Humboldt State, he still couldn't bring himself to go on campus, and he continued to have friends go to lectures for him.

Now that he had a job, Taylor felt that he needed a more conventional look, so he got his hair trimmed into a mullet. He started wearing tank-top T-shirts and ultrabaggy jeans, cut rapper style. The look helped him fit in as a checkout clerk. The C&V Market's primary products were beer, cigarettes, and lotto cards. Prostitutes came by in the morning hours, looking wrecked, and he had to keep an eye on them to make sure they didn't steal food. He began eating breakfast at Stanton's Restaurant in Eureka, a family-style trucker's restaurant on Highway 101. His usual order was bacon and eggs, hash browns fried in griddle fat, and biscuits with homemade gravy, with a couple of sausage patties on the side. By lunchtime he would be hungry again, and he grazed on frozen pizza from the shelves of the C&V Market. Soon he was buying larger and larger sizes of rapper pants.

He rented a room in a house in a neighborhood inhabited by drug addicts. There were methamphetamine labs in the mountains around

Eureka, and crystal meth—speed—was cheap. Drug dealers will sometimes cut crystal meth with battery acid. If you inject crystal meth mixed with battery acid into your bloodstream, your skin can break out into sores. Speed addicts tend to act compulsively, often picking at their sores, and the addicts are sometimes called tweakers.

Taylor thought that some of the people living in his house had a tweaky look. His diesel Rabbit passed two hundred thousand miles, and it began to die. He sold the car for fifty dollars to some rockers, people who like to do air tricks with junk cars. The rockers took the Rabbit to the Mad River in Humboldt County, where they drove it off a jump and got thirty feet of air with it, but they crashed it and left it in the Mad River.

He bought a bashed-up Pontiac Le Mans for a hundred dollars. It had doors of mismatched colors and gray fiberglass patches all over it, and the back window had been smashed out. He left it that way, figuring that if he replaced the window a tweaker would just break it looking for something valuable. Taylor himself had no use for drugs. He could get himself into another world simply by visiting a redwood park. He began driving his Pontiac down to Humboldt Redwoods State Park, up to Prairie Creek Redwoods State Park, and to Jedediah Smith Redwoods State Park—the areas of forest that had been saved by the Save-the-Redwoods League.

A WOMAN NAMED CONNI METCALF, WHO WORKED BEHIND THE COS-metics counter in a Long's Drug Store in Eureka, lived around the corner from the C&V Market. She often went to the market to buy groceries, and she'd noticed this gentle-looking, reserved man at the cash register. He seemed to be a kind person. He always had something nice to say to her, and he was actually very good-looking. She could tell that he was admiring her in a shy kind of way. A while back she had broken up with someone she thought she loved, and she was beginning to think about having a man in her life again.

One day at the cash register, she said to Taylor, "Do you have a girlfriend?"

He flushed. "Well . . . not at the moment."

"Would you like to ask me out?"

"Well—sure."

"Why don't you call me sometime?"

"All right."

"Here's my number." She jotted it down on a piece of paper and started to hand it to him.

He reddened. "That's okay. I already have it."

She smiled. "How do you know my phone number?"

"I saw it on your checks. I've wanted to call you, but I didn't. I was afraid you might think I was a stalker type."

"Oh, you're not a stalker type. You can call me, okay?"

She ended up in his apartment one night, and they became lovers. He lived in a nearly bare, single room. It had a bed, and a desk with a chair, and there were four cats living with him. He loved cats. She thought that you could tell a lot about a man by seeing how he treated animals. The walls of the room were covered with beautiful photographs of redwood trees, most of which Taylor had taken himself. On the floor was a cardboard box in which he stored some weird-looking instruments. The instruments were made mostly of plastic, string, duct tape, and plywood. He said that he was using them to measure redwoods. There were piles of maps of the North Coast, aerial photographs of redwood valleys, and copies of scientific papers about tall trees. He was a gentle lover, very protective of her. He said that he was looking for the world's tallest tree. He seemed to be a complicated man, for a grocery clerk.

AROUND CHRISTMAS 1990, MICHAEL TAYLOR BECAME FRIENDS WITH a man named Ron Hildebrant, who was working the night shift at the Eureka Post Office. Hildebrant is a soft-spoken man and a devout Christian. He had been making the rounds of the redwood parks, measuring the heights of redwoods using a homemade instrument, and he, too, had come to the conclusion that the tallest trees on earth had yet to be discovered. He showed Michael his measuring device, and Taylor began building similar gadgets.

Taylor built them out of cheap parts that he bought at hardware stores—much as he had once built stuff out of junk as a teenager. He decided to make a device called a clinometer. A clinometer works a lot like the Abney level that Paul Zahl used to discover the Tall Tree, but it's cruder. It can be used to estimate an angle from a point on the ground to the top of a tree. Taylor made his clinometer out of a plastic protractor (the kind used by elementary-school students for measuring angles), along with a piece of string, a thumbtack, and a wooden pencil. The device cost him forty-five cents.

In February 1991, Michael Taylor and Ron Hildebrant visited Humboldt Redwoods State Park on their first expedition in search of very tall redwoods. Taylor showed Hildebrant a supertall redwood that he knew about, which is now named Laurelin. They measured it with their instruments, which indicated that Laurelin was roughly 369 feet tall. They thought it might be the world's tallest tree. They suspected that it was taller than the Tall Tree had been when the National Geographic Society and measured it, but they couldn't prove it.

Taylor realized that his forty-five-cent device wasn't really accurate enough to find the tallest tree. Since he couldn't find the tallest tree, he thought he would look for very *large* redwoods instead— obese giants—and he got a measuring tape so that he could measure the girth of a fat tree. He had seen a lot of really fat trees in Prairie Creek Redwoods State Park. One day he presented himself to the rangers at the park headquarters and told them that he was interested in discovering fat redwoods; he asked the rangers if they knew of any places in the park where he should be looking for them. The rangers thought he was a woo-woo type who needed to hug a big tree, so they told him to have a look at a tourist attraction in the park called the Big Tree. It was supposed to be the largest redwood in Prairie Creek Redwoods State Park.

"The first time I saw the Big Tree, I said, 'No way is it big,'" Taylor told me. He ran a piece of string around it anyway, to measure its circumference, and he estimated its height with his plastic protractor. He was a little disappointed to find out that the Big Tree is actually really big. It's twenty-two feet in diameter at breast height, and it's 305 feet tall. Taylor, however, felt certain that there were much bigger redwoods lurking out there, Moby Dicks of the vegetable kingdom that would make the Big Tree look like a wimp. He just didn't know where to find them.

Discovery, in the case of a giant tree, doesn't necessarily mean that nobody has ever seen the tree (though this may be the case); what discovery really means is that nobody has ever understood the tree's size or measured it. According to an unwritten tradition of botany, the discoverer of a giant tree has the right to name it.

Unbeknownst to anyone else, except Conni Metcalf and Ron Hildebrant, Michael Taylor began systematically searching for the world's largest redwoods, concentrating his efforts around Prairie Creek. Along a tributary of Prairie Creek called Godwood Creek, he found a monstrous redwood that he named the Godwood Creek Giant. It is twenty-four feet across near the base—a couple of feet wider than the Big Tree, anyway—and 354 feet tall, fifty feet taller than the Big Tree.

Taylor began experimenting with better instruments. One day he was visiting a friend who had been cleaning out a storage shed, and the friend pulled out a nineteenth-century surveyor's transit made of brass. It dated from the Victorian era. The friend gave the transit to Taylor, and he cleaned it up and started using it to measure angles, so that he could estimate the height of a tree with greater precision. The Victorian brass transit was superbly accurate.

As the spring of 1991 turned into summer, Michael Taylor began bashing around in deep pockets of rain forest. Conni Metcalf saw less and less of him, and it troubled her. He got himself into nooks and notch-shaped valleys where there were no trails. Because of his experience in wilderness bushwhacking, he felt comfortable moving around the rain forest on his own. He found a redwood titan that he named Adventure. Michael Taylor's Adventure Tree is thought by botanists to be one of the world's largest and most interesting living things. He discovered and named Sir Isaac Newton, a redwood that is thought to be one of the largest in the world. One day, while he was swatting his way through a dense patch of rain forest, he discovered a hulking, rotted out, dying redwood titan that is virtually invisible in the undergrowth until you get right up to its base, when it suddenly looms like a wall. It was near a busy road, yet it had never been noticed. He showed it to Ron Hildebrant, who named it the Terex Titan, after the world's largest truck, the Terex Titan. Terex Titan Tree seems to be hollow, and it may not have any central growth rings left. It may be the oldest living redwood.

. . .

JIM TAYLOR WAS WORRIED ABOUT HIS SON, AND HE BEGAN HAVING nightmares that Michael had been shot during a robbery at the market. The neighborhood where Michael lived wasn't safe. There were guns everywhere, a lot of speed dealing, a lot of pot farming, a lot of general scumminess. Michael seemed careless of his own safety. He didn't seem to see how vulnerable he was. He didn't seem to have any sense of his own future, or any real sense that we all die one day. He didn't seem to have any perception of time in the conventional sense. Jim thought that it would be a big mistake to begin supporting Michael again—it would rob him of his independence and his dignity—but he feared for his son's life.

Jim Taylor couldn't understand what had caused his son to become an overweight grocery-store clerk with a mullet who lived in a rented bedroom in Eureka. He trusted Michael—he knew that he wasn't using drugs—but he didn't know exactly *what* he was doing. Michael had stopped calling home, and he didn't even call Jeb much. Jim decided that he had to give his son the benefit of the doubt.

ONE DAY DURING HIS EXPLORATIONS, MICHAEL TAYLOR FOUND HIMself in a dense grove of redwoods, in a place he had never been before. He was lost. He had gotten into a deep pocket of redwoods in the Fieldbrook Valley, which was rumored to hold the greatest of all the redwood titans. He was carrying a wooden tripod across his shoulder, and attached to it was his old brass surveyor's transit. He was wearing a dark wool coat with a velveteen collar, wool trousers, a white linen shirt, and stiff leather boots. In front of him stood a coast redwood that was bigger than any redwood he had ever encountered. His heart was beating like mad, and he was fumbling with his instrument and almost crying. He started to measure an angle to the redwood's top, but he couldn't see the top. The tree ran out of sight. He had discovered the Ultimate Tree.

He woke up in his bedroom, the dream fading slowly. Conni was sleeping beside him. He kept having that dream, that recurrent dream of finding the Ultimate Tree, and it made him feel sad and lost. He had been born in the wrong century, he thought. He should have been born in the nineteenth century, because then he could have found the biggest trees. There was no chance that his quest would succeed now, because the biggest trees were gone. It was as if his life's work had slipped away unfinished before he had even been born. All he could find, at best, were a few remnants of a lost world.

WHEN YOU LOOK AT A REDWOOD TREE, IT'S VERY DIFFICULT TO GET A true sense of its size. A giant redwood, in terms of sheer bulk, can be anywhere from four to twenty times larger than a merely big redwood, but it may not look much larger, at least not to most people. It's also very difficult to get a feel for the height of a redwood just by looking at it. "I can see trees pretty well," Michael Taylor said to me. He also got a feel for the existence of groves—collections of redwood giants growing together in a group. Some of the groves in Prairie Creek Redwoods State Park were famous, and had been given names, and there were signs that identified them. But there were other groves in the park that no one seemed to have noticed.

Taylor's searches took him a while. Redwood forest is choked with vegetation in many places, including thickets of barbed salmonberry canes and poison-oak vines, which, if you get into a mass of them, can give you a hospital-quality case of poison-oak rash. The land can be brutally steep, more vertical than horizontal, and so jammed with underbrush that it's almost impossible for a person to move or see anything. The terrain is sliced by narrow gorges and gulches, which may be filled with boulders and huge, fallen redwood trunks, which lie across one another like pickup sticks. You could be crawling through ferns and underbrush and pass right by a huge tree and never see it.

After about three months of exploring Prairie Creek, Taylor began to realize that a certain spot in a certain valley was particularly rich with big redwoods. The first tree that he found in the area was an incredibly large redwood with a quadruple array of huge trunks in its crown. He named the tree Atlas, because it reminded him of the Titans, the primeval deities in Greek mythology. There were other redwoods of vast size growing around it. This wasn't just another grove but something extraordinary.

Today, some botanists regard Michael Taylor's Atlas Grove as the Sistine Chapel of the world's forests, one of the greatest works of nature on earth. It is a tight jam of huge redwoods, many of them growing with their crowns almost touching. The Atlas Grove does not appear on any published map of Prairie Creek Redwoods State Park, and no signs mark the trees in the grove. The general location of the Atlas Grove is known only to a handful of botanists and their associates. As far as I can tell, just two people—both are botanists—know the names of all the redwood giants and titans that stand in the Atlas Grove and are able to point to the individual trees by name. I have been in the Atlas Grove many times. So far, I'm still uncertain of its extent.

IN 1966, THREE YEARS AFTER PAUL ZAHL DISCOVERED THE TALL TREE, a forest-soil scientist at the University of California at Berkeley named Paul J. Zinke and a graduate student of Zinke's named Alan G. Stangenberger, had begun searching for extremely tall and large redwoods in Humboldt and Del Norte counties. Zinke had found evidence in the soil that a vast flood had occurred along the Eel River about a thousand years ago—a so-called thousand-year flood—which had killed many redwoods. He wanted to find and study redwoods that had survived the flood. They had been large, tall trees then, and by now the survivors would be the largest and tallest redwoods in the forest. Zinke and Stangenberger discovered a gigantic redwood growing in the middle of Founders Grove, in Humboldt Redwoods State Park, along the Avenue of the Giants—a well-known tourist area. The two men named the tree the Dyerville Giant, after the hamlet of Dyerville, which is about a mile away from the tree.

The Dyerville Giant was obviously one of the most impressive survivors of the ancient flood. Not only was it a very massive tree, it was also incredibly tall, a skyscraper of a redwood. The Dyerville Giant was seventeen feet across at breast height above the ground. The hard part was figuring out how tall it was. The challenge of measuring a thirty-six-story-tall tree to within an accuracy of one inch is formidable. It's very difficult to see the top of a redwood from the ground, because the top is typically out of sight, buried in the canopy and screened by the crowns of other redwoods around it. The ground may be sloped, which is another problem. In addition, the top of a redwood normally consists of many bushy sprigs of foliage. It can be nearly impossible to tell which sprig is actually the top of the tree. The other sprigs are called false tops. It's only too easy to measure a false top and get a false reading. The error can be made worse by any movement of the tops in a wind.

The Dyerville Giant had a pronounced lean—it was tipping over—and that made it especially hard to measure. After trying to estimate its height using surveying equipment, and not being satisfied that they were getting the right numbers, Zinke and Stangenberger decided to try something different—they got a weather balloon. Stangenberger stood next to the tree and sent the balloon floating up beside it; the balloon was tethered to a line. Meanwhile, Zinke stood on the Avenue of the Giants, a half mile away, on top of a grass-covered roadcut called Duckett Bluff. From there he could see the top of the Dyerville Giant poking above the canopy. He watched as the balloon popped up near the Dyerville Giant and talked to Stangenberger by radio. When the balloon was hovering exactly at the top of the tree, Zinke told Stangenberger to reel it in. Stangenberger made a mark on the line and reeled it in, then he measured the length of the line.

"It was fun hauling out the weather balloon in state parks. All these tourists would watch," Al Stangenberger told me. But the scientists kept losing balloons in the dense canopy, and the lines got tangled. They gave up on weather balloons, and, in the summer of 1966, Zinke hired a professional surveyor to measure the height of the Dyerville Giant. The man used a sophisticated method that indicated that the Dyerville Giant was the world's tallest tree: it was 369.2 feet tall, or a foot and a half taller than the Tall Tree, in the Tall Trees Grove, in what would one day become Redwood National Park—ninety miles north of Humboldt Redwoods State Park (where the Dyerville Giant stood).

The establishment of a national park for redwoods was a high priority for environmentalists at the time, and the fact that the Dyerville Giant seemed to be taller than the Tall Tree may have been disturbing news to some people. The crown jewel of the national park, it was hoped, would be the world's tallest tree, the Tall Tree. If, in fact, it wasn't the tallest tree, would Congress be less interested in creating the national park? Then another tall-tree hunter named Rudolf W. Becking, who was a professor of forest ecology at Humboldt State University, paid a visit to the Dyerville Giant and estimated that it was "only" 360 feet tall, or seven feet shorter than the Tall Tree. Paul Zinke hired his surveyor to return to the Dyerville Giant and do another estimate. This time the Dyerville Giant came out at 358 feet tall. It seemed to be getting shorter every time someone measured it.

DURING THE MONTH OF MARCH 1991, AROUND THE TIME MICHAEL Taylor discovered the Atlas Grove, a series of Pacific storms passed over Northern California, bringing strong winds with heavy rains, which drenched the soil and softened it. The storms hit Founders Grove particularly hard, and four or five redwoods that stood around the Dyerville Giant were blown down. These trees were possibly around a thousand years old—they had apparently grown up around the Dyerville Giant after the flood of a thousand years ago. Their falls opened a large hole in the canopy around the Dyerville Giant, which exposed the tree to the direct force of winds for the first time in perhaps a thousand years. One of the fallen trees broke through the roots of the Dyerville Giant.

A redwood tree sits on a flat pancake of roots, spreading in all directions away from the tree. A redwood has no taproot. A taproot is a strong, vertical root, shaped like a carrot, that stabs straight down under a tree and acts as an anchor, helping to keep the tree upright. The pancake of roots under a redwood spreads out and narrows down into a fine, dense mat of threads no more than about two feet thick. These fine roots extend outward for unknown distances from the tree, perhaps a hundred yards or more. They eventually merge with the threadlike roots of other redwoods, forming a tangled mat of roots. The roots of a redwood forest resemble a pad made of felt. The pad seems to support all the redwoods that are in a stand; they are all anchored by the common mat. The Dyerville Giant was a tower of wood held upright by a carpet of threads. The tree falling next to it had smashed through the carpet.

In the days following the storm, the Dyerville Giant seemed to be leaning more than usual. No one thought much of it. On Sunday, March 24, 1991, a park ranger reported that the Dyerville Giant was leaning over in a way that seemed very peculiar to him.

The next morning, Monday, March 25, was calm and windless. At about 7:45 a.m., a man who lived in a house overlooking the Eel River, a mile to the south of Founders Grove and on the far side of a ridge, was awakened by a deep roaring sound, followed by a boom that shook his house. He ran outdoors, thinking that a freight train had derailed on the train line that ran below his house. The tracks were empty. The Dyerville Giant had gone down. Its death was reported on the A.P. newswire.

THE FRESH GRAVE OF A COAST REDWOOD IS KNOWN AS A DETONATION zone. The day after the fall of the Dyerville Giant, Michael Taylor went to see it. Crowds of people were standing around the broken hulk of the tree. Five other big redwoods had gone down, and wreckage was strewn everywhere. There was a crater forty feet across, where the tree's root mass had been torn out of the ground as the tree toppled over. The tipped-up root mass extended at least thirty feet into the air. The trunk was a ridged cylinder, far taller than the people staring up at it. The people were quiet, and many of them seemed to be in awe. Someone had placed a bouquet of flowers near the root crater.

Taylor noticed some damage far up in a redwood that grew near the Dyerville Giant. He could see what had happened: the top of the Dyerville Giant had raked through this redwood as it fell, stripping the bark off it and leaving wounds that to Michael looked like cat scratches. Taylor named that tree the Cat Scratch Tree. Ron Hildebrant visited the site, too, and with their measuring instruments they found that the highest scratches on the Cat Scratch Tree were 108 feet above the ground. They calculated that in order for its top to strike the Cat Scratch Tree 108 feet above the ground, the Dyerville Giant must have been at least 370 feet tall. So it really had been the world's tallest tree. They were the only ones who knew it.

. . .

On a gray morning in February 1993, two years after he first began to explore the redwood groves, Michael Taylor decided to try to get in touch with Professor Paul Zinke, the man who had discovered the Dyerville Giant. Zinke was an emeritus professor at the Center for Forestry at U.C. Berkeley. Taylor wondered if Zinke had found other world-record redwoods—redwoods that only the old professor knew about.

INTO THE GROVES
OF THE SUN

ICHAEL TAYLOR INTRODUCED HIMSELF TO PROFESSOR
Zinke on the phone and said that he was searching for
giant redwoods. "How did you discover the Dyerville
Giant?" he asked.

Zinke had a gravelly voice, and he was polite. ("He spoke to me
as I think he would speak to another scientist," Taylor told me. "I
didn't tell him I was a grocery clerk.") "It's obvious that the tallest
redwoods haven't been found," he went on to Zinke. "Where do you
think I should be looking?"

Zinke advised him to begin his search in Humboldt Redwoods
State Park. He thought Taylor would do well to concentrate on the
flat areas at the bottoms of valleys in the park.

"How do you spot the tallest trees?" Taylor asked.

Zinke advised Taylor to hike up along the sides of the mountains
to the old Indian prairies, the grasslands in the forest, where he could
get a clear view. From such a vantage point, he would see redwood
tops sticking up. These would be very tall trees. "Just a minute,"
Zinke said abruptly, and he left the line.

Taylor waited. Minutes ticked by. He heard clattering and clunking sounds—it sounded like the professor was digging around in filing cabinets. Eventually, Zinke returned and explained to Taylor that he had found some of his old data from the sixties. There were three giant redwoods to look for. They were the Three Peas in a Pod, he said. In the data, they were numbered 12, 13, and 14. He had nailed metal tags to these trees. "If you find any tags with those numbers on them, you've found the Peas in a Pod," Professor Zinke said. They could easily be the world's tallest trees, he added, and politely bid goodbye.

MICHAEL TAYLOR BEGAN TO EXPLORE HUMBOLDT REDWOODS STATE Park. He was able to measure one or two trees a day, provided he got a whole day off at the grocery store.

He smashed his way through thick, almost impassable places. He climbed up to grassy overlooks and stared across the valleys. He wondered if the Peas in a Pod had fallen. Standing on an overlook, he spotted tall emergers sticking up here and there. They were usually a mile or two away. Even though he could see a tree's top, he still had no idea which trunk it belonged to on the ground. It was easy to see the tops of the tallest trees. Finding their bases was the hard part.

He would do this by taking three compass bearings on the top of the tree, from three different locations. Then he would go down into the groves and hike along the three bearing lines, through deep underbrush. Eventually, if he was lucky, the three lines would converge at a spot where there was the lower trunk of a seemingly tall redwood. At that point, he would set up his brass surveyor's transit and try to get an estimate of its height.

In 1994, Taylor inherited twenty thousand dollars from his great-grandmother, Florence Taylor, who had died at the age of ninety-seven. The money seemed like a fortune to him. He used some of it to pay his college bills and invested the rest in nearly full-time redwood exploration. Conni Metcalf also began to support him financially, helping him with her earnings at the cosmetic counter at Long's Drug

Store. She was making scientific grants to her boyfriend at a time when he couldn't possibly hope to get any money from the National Science Foundation or the National Geographic Society.

The National Geographic Society had once made a big deal of the way it had allegedly explored the redwood forests, but in fact the Society had totally dropped the ball. Executives in Washington, D.C., seemed unaware of the fact that one of the most important ecosystems in North America remained unexplored at the most basic level, the level of a map. They would never have given money to someone like Michael Taylor, anyway, since the highlights of his résumé were door-to-door knife sales and mini-grocery management, and his chief collaborator was a night-shift worker in the post office.

IT IS THE RIGHT AND THE PRIVILEGE OF AN EXPLORER TO AWARD names to things. The tallest trees on earth are redwoods in a class of height above 350 feet. A redwood that's more than 350 feet tall is a rare organism. There aren't many of them. As he picked his way through the valleys of Humboldt Redwoods, Michael Taylor began finding redwoods of world-class height: redwoods that he and Hildebrant named Graywacke (after the gray sand that gives the beaches of the North Coast their color), Gray Poison (an old, gray redwood, covered with poison-oak vines), Thunderbolt (it had been struck by lighting). Taylor found Crossroads, and he found Paradox. Sometimes collaborating with Hildebrant, but often exploring the forest alone, Taylor discovered other supertall redwoods that ended up being named Springing Buck, Laura Mahan, Bamboozle, Logjam, Thor, Bushy Toe, and Warm Winds. Taylor and Hildebrant named a 362-foot tree Harriett Weaver, after an author of books about redwoods. One day Taylor found a soaring redwood, 363 feet tall, which he and Hildebrant named Alice Rhodes, after Hildebrant's grandmother. ("She loved redwoods," Taylor explains.) Taylor was making important discoveries in the sense that he was beginning to identify and measure the dominant living things in the forest—important organisms in an important ecosystem.

He destroyed several pairs of expensive hiking boots. No matter how good the boots were supposed to be, they would fall apart after a few muddy bushwhacks. Finally, he went to a Kmart store and bought a half-dozen pairs of the cheapest running shoes, for five dollars a pair. He would wear them for a few trips into the forest and throw them away. He ended up throwing shoes out once a week.

Meanwhile, the Le Mans was running through his money at an alarming rate. He would put two or three dollars' worth of gas in it at a time, whatever he had in his pocket, and he usually ran the car close to empty.

He was worried that he wasn't giving enough time to Conni Metcalf. He would spend all day in the groves and return covered with mud, stinking of sweat, without even a few dollars in his pocket, because he'd spent it all on gas. The soles of his running shoes would be peeling off, and he talked endlessly about trees. He would take a shower, put on clean clothes, and they would eat dinner in the apartment. Usually she cooked. Michael would tell her how his search for giant trees was going. She would tell him about whatever had happened at the cosmetics counter or around town. They would make love, and he would fall asleep in her arms. But then he would wake up. His restlessness at night disturbed her. Sometimes he had nightmares, too—dreams that he was falling, for his fear of heights was worse than ever. Then he would get up and pace the room, and sit down at his desk, poring over maps of the North Coast.

CONNI METCALF THOUGHT THAT MICHAEL TAYLOR WAS STRANGE. He loved trees, almost too much. When she first met him, she'd seen him as a cute, hunky guy, a pickup at the cash register. Now, as she weighed her feelings for him, she realized that it might be his passion for trees that was making her feel strongly about him, although it also drove her crazy. When he drifted into her apartment late at night or left at the crack of dawn, she would say to him, "I'm a tree widow."

He was a man who could find beauty in the small, hidden places that still existed on earth, the lost places that nobody had ever noticed. Michael was the stubbornest person she had ever known. He bore a resemblance to the great explorers who had lived in earlier ages, and had been convinced that there was something wonderful still to be found on the earth.

THE SKYWALKERS

IN THE SUMMER OF 1988, STEVE SILLETT HAD BEGUN DOING RE-
search for his senior thesis at Reed College. He planned to study
lichens in the tops of old-growth Douglas-firs. Reed College students were allowed to do research in a tract of trees along the Sandy
River, east of Portland. There was a cabin on the land for students,
and a man named Sam Diack lived on the property.

Diack often helped Sillett climb the trees. Sillett was climbing
them the way he had learned from Bill Denison. He used a slingshot
to fire a fishing line into the branches, and then he used the fishing line
to pull a rope up over a branch and down to the ground. He would
climb up using Jumar ascenders, and then, when he got to the top of
the rope, he would climb like a rock climber up to the top of the tree.
Sam Diack stood on the ground holding a belay rope that was attached to Sillett. A belay rope is a safety rope, anchored loosely somewhere below the climber. If a climber falls on a belay rope, he can
sometimes fall quite a distance before the rope pulls tight. Rock
climbers call this taking a whipper. Sillett once fell from near the top
of a Douglas-fir and took a huge whipper. Sam Diack saved his life by

throwing his weight on the belay rope and stopping Sillett's fall. It tore the muscles in Diack's back.

That summer, when he wasn't climbing trees along the Sandy River, Sillett occasionally stayed with a classmate named Joe Jeral, who lived in a house off the Reed campus. Jeral had a housemate, Amanda LeBrun, who had just finished her freshman year at Reed. Earlier that year, she had braided Steve's hair into a French braid for him, but they had hardly spoken to each other. Their first meeting was virtually wordless and delicately erotic. Months later, still without ever having had much of a conversation, they made love.

Amanda LeBrun was a soft-spoken person, with an honest, direct manner. She had long curly brown hair, freckles, and green eyes. She was studying English and Spanish poetry at Reed, and was writing poetry. She found Sillett's passion for trees intellectually seductive, and she began going out to the Sandy River to watch him climb. She became his first girlfriend.

One day Sillett decided to throw LeBrun a physical challenge. "This was my modus operandi, how I gave people a litmus test for friendship," Sillett said to me. He invited her to climb a 250-foot Douglas-fir. She accepted his challenge, and she made it to the top without much difficulty.

Amanda LeBrun came from a wealthy family in Dallas, and she had done a lot of sailing in yachts in the Caribbean and among the Greek Islands. She had developed a feel for ropes and knots, and good instincts for what to do when a situation is dangerous and changing fast. While climbing trees with Sillett, she discovered that he had a fear of heights, and that he didn't deal well with stress in climbing. Once he dropped his rope while he was climbing a Douglas-fir, and got stuck in the tree 150 feet above ground, with no way to get down.

"Steve is a very commanding presence, but if he isn't in control of a situation he freaks out," LeBrun said to me. "He absolutely lost his shit, to the point of being almost hysterical."

Standing on the ground below, she urged her boyfriend to stay calm while she figured out a way to get him down from the tree. She had him send down a length of nylon cord that he was carrying, and she tied his rope to it; he then pulled the rope back up into the tree, and she tied a sailing knot in it to keep it anchored as he came down. Their relationship became extremely passionate after that incident.

Sillett graduated from Reed College in 1989, and he went on to graduate school at the University of Florida, in Gainesville, where he studied forest-canopy biology with Nalini Nadkarni, one of the ecologists who had become interested in climbing trees in the late 1970s while exploring the tropical-forest canopy in Costa Rica.

Amanda LeBrun wanted to have more time to write poetry, so she took a leave of absence from Reed and went to Florida to be with Sillett. They spent the summer of 1990 living in a converted chicken coop on a mountain in Costa Rica's Monteverde Cloud Forest, where they climbed giant fig trees and collected samples of mosses and liverworts. LeBrun became Sillett's field assistant. She wrote a little, but she found herself mainly working in the trees with her boyfriend.

Sillett discovered that he didn't like doing research in tropical forests. There was too much biodiversity, a profusion of different species everywhere one looked. It took him and other biologists a total of four years, ultimately, to identify 127 different mosses and liverworts that he and LeBrun had collected from just six fig trees in Costa Rica. The work frustrated him, and he began to feel that he wasn't getting any closer to the big discoveries that he wanted to make about forests—all he was doing was identifying scraps of green stuff. "Also," LeBrun explained to me, "the trees in the tropics weren't tall enough for Steve."

Sillett wanted to climb in tall temperate rain forests. He left Florida State and went to Oregon State University, in Corvallis, to get a Ph.D. in botany and to do research in the canopy in Oregon. LeBrun returned to Reed College and finally graduated, and they began to live together. She got a job as a schoolteacher in a Montessori school and helped to support him. By this time, he had finally gotten a driver's license, but now he needed transportation. LeBrun bought a pickup truck, and he used it to ferry climbing gear to various patches of forest around Oregon, where he studied lichens of the canopy. They were married in the summer of 1993, in a civil-mystical outdoor wedding in a state park that featured Frisbees, loose dogs running around, and people swimming nude in a river.

Bill Denison, who had been the first explorer of the Pacific Northwest canopy, came to the wedding. By then, he had retired and was getting on in years. Denison was a bearded, wiry man, with a powerful voice and a booming laugh, and he wore black-rimmed spectacles. He had become incredibly fond of Steve Sillett, and had gotten to know and like Amanda LeBrun. The couple invited honored guests to say a few words before they exchanged their vows, and Denison got up and, in a loud voice, spoke about partners who climb together for life. "Marriage is a rope you tie between you," he said. "It's like a rope that joins two climbing partners and keeps them from falling. Marriage is about rope management. You have to take care to avoid knots and snarls in the rope that joins you together. You can't keep the rope too tight, but you can't let it get too loose, either. Each of you has to give your partner enough slack for freedom of movement, so that you both can reach the top together." His voice broke, and tears were running from behind his spectacles.

Sillett and LeBrun moved into a small house in the town of Philomath, near Corvallis. The house was off the grid—no electricity, a rainwater-collection system, and a woodstove. She taught sixth grade in a middle school. Eventually, they bought a house in Philomath. Meanwhile, he began to experiment with ways of ascending tall trees. For a while, he tried to climb trees using a method used by loggers and telephone-pole climbers.

Professional loggers had long been climbing up the trunks of redwoods, Douglas-firs, and other tall conifers. They were called high climbers, and they had mastered a technique known as spur climbing. It works like this. The climber wears boots that have long steel spikes—climbing spurs—attached to them. He wraps a stiff, heavy rope known as a flipline around the trunk of a tree, holds each end of the flipline, plants his spurs in the tree, and braces himself against the flipline. Then he walks up the tree with his spurs, flipping the rope upward along the trunk as he climbs. Utility workers also use this technique to climb up telephone poles.

When a high climber has spur-climbed up a redwood to around ninety feet above the ground, he wraps a steel cable around the trunk of the tree. After the climber sets the cable, he descends to the ground, walking down the tree using his spikes. A logging crew then cuts a notch in the tree at its base with chainsaws. A truck with a winch pulls the cable, and this, together with the notch, directs the tree's fall in the desired direction.

ARBORISTS PRUNE TREES, CARE FOR THEM, AND CUT THEM DOWN when necessary. Arborists were once called tree surgeons. Many arborists don't climb trees; in fact, if they own the business they may be somewhat older and not in shape for climbing. They often hire younger men and women, who are sometimes called climbing grunts and are paid an hourly or a daily rate to do the aerial work in trees. Aerial work is dangerous, and the fatality rate for tree-climbing grunts is one of the highest in any industry.

Arborists and climbing grunts have developed a sophisticated method for climbing trees. The arborist climbing technique, as it is called, involves the use of special ropes, sliding knots, and a safety harness called the tree-climbing saddle. Arborist ropes are thick, strong, and soft, and they're gentle on the trees. An arborist normally wears soft-soled boots, since the spiked boots worn by pole climbers and loggers can cause severe damage to a living tree. The arborist climbing technique is always evolving. Tree climbers tend to have individual climbing styles, and they have a habit of tinkering with their climbing gear and experimenting with new gadgets.

The better professional arborists are certified by the International Society of Arboriculture. The I.S.A. sponsors get-togethers called jamborees, where members meet and exchange business ideas and tree-care methods. The I.S.A. jamborees also attract grunts, who participate in climbing competitions. The leading I.S.A. champion tree climbers are top international athletes, and their abilities as climbers are legendary within their field, though they're virtually unknown outside that tiny world.

In the mid-1980s, two grunts from Portland, Kevin Hillery and Douglas Wallower, began climbing Douglas-firs, for pleasure, in the Carbon River rain forest, a stand of old conifers that fills a valley near Mount Rainier, in Washington. Hillery and Wallower were both working as climbers for the same tree-care company in Portland, and they had become friends.

Kevin Hillery was a tall, thin, rangy man, with a powerful-looking body, and he was an I.S.A. champion climber. His climbing partner, Douglas Wallower, was smaller, and he had a quick, squirrel-like way of moving his body. Wallower was, if anything, an even better climber than Hillery, though he was a loner and never competed in climbing races. "I'm a recluse, with more than a touch of misanthropy," Wallower said to me.

Both men had a highly prized tree-climbing style that tree climbers refer to as fluid movement. They climbed trees without visible effort, gliding on ropes upward and through the branches with a dexterity that seemed to question the existence of gravity.

The oldest Douglas-firs in the Carbon River rain forest are around 550 years old. There has been no major fire in the forest since the fifteenth century. Some of the trees in the Carbon River forest have grown to be ten feet in diameter at breast height, and the tallest parts of the canopy are between 280 and 300 feet above the ground. Hillery and Wallower became convinced that the world's tallest Douglas-fir might exist, hidden and undiscovered, somewhere in the Carbon River rain forest. This Douglas-fir would be more than 310 feet tall, they thought, and they referred to their hypothetical tree as the Holy Grail. They began to search for it, systematically climbing the tallest trees in the forest and measuring them with a fishing line, which they ran down along the side of the tree from its top on a lead sinker. They were joined in their search for the Holy Grail by a friend of theirs, Jon Shaffer, who lived in a little house on Wallower's property, and who had fallen in love with tree climbing and had been spending his time doing mostly that. "Nobody else, I mean nobody, was climbing in those trees when Kevin and Jon and I started out," Wallower recalled to me. (They didn't know Bill Denison, who had pretty much retired, anyway.)

At first, they had tremendous difficulty getting a climbing rope over a branch of the taller Douglas-firs. The lowest branch on a mature Douglas-fir could easily be 150 feet above the ground. The method they developed for dealing with this involved one of the men climbing a small tree, anchoring himself to it with a rope, and then swinging himself like Tarzan into a taller tree. He'd climb that one and swing himself into a still taller tree, until he'd reached the lowest branch of the tree he really wanted to climb. Then he'd let a rope down and his partners would come up, and they would climb to the top and measure the tree with the fishing line. Mainly, though, they played in the high canopy, swinging from tree to tree, often out of sight of the ground. "If you spend any time climbing in an old-growth forest, you will be changed for life," Wallower said.

Kevin Hillery was watching a local television show called *Oregon Field Guide* one day when he saw footage of a man he didn't recognize climbing a tree using spurred boots and a flipline. It made him mad, because he felt that this was very damaging to the trees. The man on the show said that he'd learned how to climb from a young scientist named Steve Sillett. Who was this Steve Sillett jackass, who knew nothing about climbing except how to kill trees with spiked boots? Hillery got Sillett's home phone number and dialed it. Sillett came on the line.

They chatted about research on Douglas-firs. Then Hillery asked Sillett about his spur-climbing method. "Anybody who needs to use spurs to climb a tree has no business climbing trees," Hillery said.

"Who the hell are you?" Sillett said.

"You're kind of a wuss, aren't you?" Hillery answered. "You're spiking your way up trees like a wimp logger. Loggers are wimps. Don't you know there's a better way to get up a tree?"

"What are you talking about? I know how to climb trees."

"Haven't you ever heard of arborists?" Hillery said, and he asked Sillett to meet him at the Carbon River.

ARBORIST-STYLE CLIMBING IS ALL ABOUT ROPES AND KNOTS. Climbers get around in trees by tossing the end of a rope over an anchor point—a strong branch or a V-shaped crotch. Once you get the rope passed over an anchor point, you have to get the end of it back to you. This is often difficult to do, and there are various tricks for retrieving the end of a rope that's a distance away from the climber. When you do get the end back, the rope is formed into a long loop, or noose, over the anchor point, and the noose is tied with a sliding knot. The sliding knot looks a little bit like a hangman's knot, though it functions in a different way. By sliding the knot, a tree climber can shorten or lengthen the loop of rope over the anchor point, thereby moving upward or downward. A climber can also lock the sliding knot and hang motionless.

The loop of rope is attached to the climber's saddle (harness) with a carabiner, a strong clip made of aluminum or steel. A tree-climbing saddle looks something like a rock-climbing harness, except that it's thicker and has more padding, and typically has all sorts of tool pouches and devices attached to it—special gadgets that are helpful for getting around in trees.

Wearing a saddle and suspended from the loop of rope, you can turn your body horizontally and plant your feet on the tree and walk up the trunk. This is called trunkwalking. You can kick off and swing, Tarzan-style. You can suspend yourself in the air between two anchor points, hanging in space at the lower point of a V of ropes. A skilled tree climber can travel horizontally or at diagonals through the crown of a tree while he's hanging in midair, and not even touch the tree with his body. This is called skywalking.

Tree climbing in the arborist style is not at all like rock climbing. Tree climbers are virtually always suspended from a rope that is anchored above them, and the rope is taut—it's holding some or all of the climber's weight. Rock climbers move upward over a vertical surface of stone by using their hands and feet to obtain friction and support. They aren't suspended from taut ropes (except sometimes in the type of rock climbing called aid climbing). A rock climber advances upward while a safety rope, trailing loosely, is held by someone stationed below. The rope runs through a carabiner that's attached to an anchor in the cliff face, a chock or a cam. The rock climber typically places these anchors as he climbs upward. The rope is there in case the climber loses his grip on the stone and falls. In tree climbing, the rope is used as the main tool for gaining height and for moving around. The bark of a tree is crumbly and soft, and a climber can't get any kind of secure grip on it with his hands and feet.

Tree climbers who are engaged in arborist-style climbing often walk lightly along branches, keeping most of their weight suspended on the loops of rope and very little weight on the branch. They call this branchwalking. A skilled tree climber can branchwalk along a branch that's no thicker than a gardening stake without breaking the branch. A tree climber can turn sideways or even upside down in the air, hanging from ropes—that latter move is called doing a bat hang. Tree climbing in the arborist style can be a kind of aerial ballet. It is a slow ballet, to be sure, for tree climbing never goes rapidly, except in an emergency. Accidents in trees happen instantaneously and without warning, and typically end in death.

"I once saw a guy die in a fall," Douglas Wallower said to me one day. "The guy was giving a safety demonstration and he made a mistake. He fell only fifteen feet, but he broke his neck in front of us and died. It was ironic. If you become a tree worker, you get an average of about five years until you have a major injury, I think. If you're careful, you'll be safe. If you take risks, you'll be smashed."

STEVE SILLETT DIDN'T BRING HIS SPUR-CLIMBING GEAR TO THE CAR-bon River rain forest, because he suspected that the grunts would laugh at him if he tried to wear it. "In my opinion, you can't climb trees worth a lick," Hillery said to him. "That's not *climbing*, what you do, Steve. Look at how much motion we have in the canopy on these ropes."

Douglas Wallower gave a demonstration. He put on a climbing saddle and then free-climbed a small hemlock tree using his hands and feet like a monkey, dragging a rope with him as he climbed up. Once Wallower had got into the top of the hemlock, he passed the rope over a branch and suspended himself from a loop of the rope, hanging in midair. Then he descended using a sliding knot, lengthening the loop until he was dangling on a fifty-foot-long pendulum. He kicked off, swinging through the air into the branches of a taller tree that stood nearby. As he swung into the tree, he grabbed the branches and anchored himself with the other end of his rope. He untied the loop of rope in the first tree, gave it a good pull, and when it fell free he pulled it over to him. Then he climbed up to the top of the second tree, anchored the rope to another branch, and once again made himself into a pendulum. He swung from tree to tree, going higher and higher. He called these motions his Rocky-the-Flying-Squirrel moves.

If Tarzan swings didn't get Wallower where he wanted to go quickly enough, he would simply take a flying leap into the next tree. He did this by making a loop of rope over a strong branch near the top of the tree. He would slack out the loop, making it long and loose. Then he would jump out of the tree like a skydiver, with the rope trailing after him, and he'd land in the target tree, grabbing the branches. The rope was there only to break his fall and save his life if, by chance, he missed the jump.

That day, as Sillett watched, Wallower performed several Rocky-the-Flying-Squirrel moves, swinging out of a hemlock into a big Douglas-fir, and then into an even bigger Douglas-fir. When he had landed safely in the branches, he shouted down at Sillett, "What we're doing, Steve, is vastly superior in every aspect—in speed, in safety, and in pleasure."

"Steve is scary bright, and I love the guy, but he needs to have his balloon pricked once in a while," Wallower said to me. "We were light-years ahead of him in climbing technique. We just smoked him."

Sillett understood what he was seeing. It was full, three-dimensional movement of the human body through tree space, movement across and through the high canopy—pure magic in the canopy, and it took his breath away. He watched how the climbers balanced their bodies as they moved along branches, how they walked along thin branches that should have broken under their weight but didn't, because the climbers were suspended on ropes attached to anchor points above them. He saw how they pivoted and swung in the air. He began to ache to learn how to skywalk. The canopy scientists who had been climbing trees with ropes had just gone straight up and down the rope. They had known nothing about skywalking. They hadn't been able to actually move through a tree. The grunts were showing him a secret door that led upward into the unexplored forests above the earth.

Sillett began buying arborist climbing gear and practicing with it in the forests of the Middle Santiam River, where he was doing his dissertation work. He learned the tree workers' language. They have certain signal words that they use when they're shouting between the top of a tree and the ground. "Headache" means that an object, such as a tool or a broken branch, is falling out of a tree. "Clear" means that all is well. Professional tree climbers refer to falling to one's death from a tree as cratering. They also refer to it as taking a dive into a dirt nap. Steve Sillett climbed often with Kevin Hillery, Douglas Wallower, and Jon Shaffer, joining them in their quest for the Holy Grail.

2

THE FALL OF
TELPERION

HEADACHE

ONE DAY IN 1991, A MAN NAMED TRUMAN JAMES, WHO WAS a tree worker employed by the National Park Service, was using spurred boots to climb a tall, straight sugar pine in Kings Canyon National Park, in the Sierra Nevada. Truman James was a Native American who was widely respected for his exploits as a tree climber. The sugar pine that James was climbing that day grew near a campground in the park, and some of its branches were infected with dwarf mistletoe, an invasive pest. Several other tree workers were climbing in the campground clearing trees of mistletoe. Truman James ascended to the top of the tree with spurred boots and a flipline, and, once there, he set a rope and began to rappel down the tree, pulling mistletoe out of the tree as he went along. He moved easily down through and around the branches in a relaxed way, with no visible effort.

As he rappelled along, he used a type of sliding friction knot called a prusik knot to get himself down the rope. Sugar pines tend to ooze large amounts of sticky sap, and when the knot became gummed up with sap he decided to clean it. He paused and sat on a branch 130 feet above the ground and stabbed his spurs into the tree to get a good grip. Then he braced one foot on a small, feathery branch for balance. It was an epicormic branch, a little branch sticking out of the bark— the type of branch that some trees shed like dog's hair. He untied the knot so that he could clean it. He was no longer attached to the tree with anything except the spurs on his boots.

Suddenly the little branch broke under his foot. Probably it had been getting ready to go. This made him teeter on the larger branch, and he kicked backward, lost his balance, and fell. He shouted, "Oh!" It was a loud shout, and some of the other tree workers turned and looked. They saw him falling along the tree, cartwheeling and hitting four or five large limbs on the way down, and he landed on the ground with a sound like a thunderclap. His co-workers rushed to help him.

To their surprise, he sat up. Amazingly, he was alive, though his face was bright red, and he was cursing angrily. He started to stand up, and two men pushed him back down, and finally they had to put their full weight on him in order to get him to lie back. He lifted them off his body and started flailing his arms. His voice began to sound gurgly. His eyes got a glazed look, and he suddenly collapsed and fell to the ground, and died not long afterward. The impact had burst his aorta. His heart pumped his lungs full of blood and stopped.

On December 21, 1993, Steve Sillett, Kevin Hillery, Douglas Wallower, and Jon Shaffer gathered at Wallower's place on the outskirts of Portland for a party to celebrate the winter solstice. Amanda LeBrun was immersed in teaching responsibilities, and she didn't go to the party. Wallower's property was in the hills above the city, and near it was a stand of Douglas-firs that were about 120 feet tall. Their plan was to climb the firs, watch the sun go down from the treetops, and then descend and drink beer from a keg and eat a potluck dinner with some friends.

The day was growing old when Kevin Hillery got a rope anchored in one of the firs. He climbed the tree, and when he got near the top he threw a rope into a taller fir nearby and skywalked into it. Then the other men climbed up. Steve Sillett free-climbed, not bothering to use a rope. They sat in the branches of the two firs and watched the sun descend below the horizon, on the shortest day of the year. It was a clear evening, and it got very cold. As darkness came on, the climbers began to descend out of the trees, one by one.

Since it was getting dark, Sillett wanted to use a rope to get down. "You can descend on my rope," Hillery said to him. Sillett was the next to the last person to descend, leaving Kevin Hillery alone in the top of one of the Douglas-firs. By then, stars were coming out. While the others began packing up their ropes, standing around at the base of the trees, Hillery prepared to descend along the rope down which Sillett had just rappelled. He decided to re-anchor the rope at a place that had a clearer path to the ground, and he went down a distance and clipped a safety rope around the trunk of the fir, attaching himself to the tree securely. Then he prepared to place the main rope over a new anchor point. It was much darker in the Douglas-fir canopy than it was in the fields and meadows around the stand of trees. Hillery thought that it was adult dark but not kid dark. Adult dark is when the adults tell you that you have to stop throwing the ball around and come indoors. Kid dark is so dark that you can't see the ball anymore, and you have to stop playing.

It was adult dark, but not kid dark, and Hillery could see branches of trees all around him. He gathered up one end of his rope and threw it toward two strong branches coming out of the trunk. The rope snaked upward. He felt it catch over the branches. He got the rope looped over them and clipped it to his saddle. Then he un-clipped his safety line and swung out into space alongside the trunk of the tree, hanging from the branches.

The "branches" tore off, and he went into free fall. His rope had missed the secure branches and had become entangled around an epi-cormic spray—a fan of feathery epicormic branches as thin as pencils, sprouting from the bark of the tree. They tore out, and he fell. He was ninety-six feet above the ground, almost twice the height of the red-line, the line of death.

As he started to fall, Kevin Hillery called out, "Headache," to in-dicate that an object was falling from the tree. Then, as he accelerated beyond the redline, he maintained professional silence.

CRATER

WHEN JON SHAFFER HEARD KEVIN HILLERY CALL OUT, "Headache," he stepped backward from the base of the tree, thinking that Hillery had dropped something and wanting to get away from whatever was falling. The others looked up. They saw Kevin Hillery falling upside down through the air, backward, having started his fall nearly a hundred feet above them. By then he had tipped over, like a weighted arrow. He was upside down, falling head down and backwards, and was sweeping his hands through the air, clawing at it, as if he was trying to get a grip on the air and pull his head up higher than his feet. There was a rising hiss. It was air running past his clothing, growing louder as he sped up. He was getting into the range of terminal velocity. Over the sound, Sillett screamed, "*Kevin . . . !*"

Since Hillery was falling backward and looking up, he would have seen the forest canopy receding from him. He must have been ransacking his mental library of climbing knowledge, hoping to find some gesture of black art that might help him survive the unsurvivable. He was going to land headfirst. Douglas Wallower had always said that a head landing was the best way to come down in a long fall, because there was absolutely no chance you'd end up a quadriplegic.

They saw Hillery twist his body around. He rolled over, still clawing the air, and turned his face downward, looking at the ground. The ground was rushing toward him at a speed that would reach ninety miles an hour at the moment of impact. If he saw the ground coming at him, the rush would have been fast enough to create tunnel vision—a radiant smearing-out of the edges of the view, with a circle in focus at the convergent center. Maybe he couldn't tear his eyes from the spot as it came up. On the other hand, maybe he didn't see anything, maybe he went inside himself, trying to escape inward and away from what was coming. He still didn't scream.

As the ground came up, he put out his left arm and reached across his chest with his hand held palm outward, as if he was trying to ward off the ground.

HILLERY LANDED DIRECTLY AT JON SHAFFER'S FEET. THERE WAS A deep, wet boom when he hit, mixed with a whooshing sound. It came from all the air being driven out of Hillery's lungs with explosive force. His body compressed the soil, and Douglas Wallower was amazed to see a beautiful sparkling radiance of light expanding in a cloud around Kevin Hillery, rising out of his body and spreading away in all directions. Wallower believed, he said later, that he had seen Hillery's soul leaving his body.

Hillery ended up lying on his side in a depression that his body had made in the ground. He had literally cratered. He was motionless, and looked as if he were asleep.

Sillett let out a cry. Jon Shaffer bent down and saw that Hillery's eyes were shut, and that he wasn't breathing. He had gone into breathing arrest. Shaffer got ready to give him CPR in an effort to restart his breathing, when, suddenly, Hillery's lungs reinflated with a huge, hoarse, weird sound.

Then his limbs began shaking and twitching, and he started having tremors or small seizures. The tremors passed, and his eyes opened. He seemed to be looking at something in the distance. "The wind rushing through my hair . . ." he said.

Douglas Wallower took off running down the hill toward his house.

"It's okay," Hillery said. "Nothing hurts."

Shaffer knelt over Hillery and ordered him to stay still. "You're not moving, buddy," he said. If Hillery's back or neck had been broken by the fall, any movement of his body could cut his spinal cord. There was no sign of blood on his exposed skin or clothing. There was no way of knowing if he was flooding himself with internal hemorrhages.

Hillery started apologizing. "I'm really sorry you had to see that. I don't think I'm hurt," he gurgled.

Sillett noticed the sound in his voice. "I can't stand listening to his gurgling," he said. This agitated Hillery, who now heard the gurgling himself. He knew that he had broken his ribs on the impact, and he was afraid that the broken ribs had punctured his lungs, causing them to fill with blood. He gurgled that he was starting to feel pain. It was in the deepest places inside his body, the worst pain ever, and it was getting worse and worse. He began to make small moaning noises, lying on his side in the hole. Shaffer kept his hands on Hillery, talking softly to him, trying to keep him from losing consciousness and going into another breathing arrest.

Wallower arrived at his house and called 911. When the operator answered, he began vomiting. The operator waited. When he regained his ability to speak, Wallower told the operator that a man had fallen a hundred feet out of a tree but was still breathing. He asked for help.

At the base of the tree, Kevin Hillery remained motionless. He seemed to be able to move his legs a little, and they concluded that his spinal cord had not been severed. He was in light shock, not deep shock. He understood this. He said that he was glad he was not in deep shock.

Sillett was frightened by the appearance of Hillery's left arm, and he pointed it out to the others. Hillery lifted up his arm to look at it. The wrist was shattered, like jelly, and his hand flopped over backward. He said it didn't hurt.

Ten minutes after the fall, two medical technicians arrived, running on foot. They lived in the neighborhood and had heard the emergency call on their scanner radios. They knelt down by Hillery and inspected him, but they didn't try to move him. "I've seen worse," one of them said out of Hillery's hearing. They were worried, most of all, about internal bleeding. They began snipping his outer clothing off with scissors.

"I don't need a helicopter," Hillery said.

An ambulance arrived and stopped about two hundred yards away, and more people ran to the scene, carrying a stretcher board. They began to move Hillery onto a board, and that was when the pain really hit him. It made him delirious, and he started babbling, slipping in and out of consciousness. They got him strapped to the board. A helicopter from the Life Flight Network touched down on a nearby road, and the team carried Hillery to it, and he was loaded into the helicopter and it lifted off and flew toward Portland. The bouncing and vibration of the helicopter caused Hillery to experience almost unbearable agony. The trauma team couldn't give him painkillers, because they were afraid that the drugs might stop his breathing or his heart. A minute or two after the chopper touched down on the roof of the Oregon Health & Science University Hospital, a team in the emergency room began working on him. A technician inserted a catheter into him. "Just what the hell are you doing?" Hillery yelled and grabbed the man's hand and wouldn't let go. They shot him full of painkillers, and he went into a dream.

In the intensive-care unit during the night, Kevin Hillery remained awake but hazy. He didn't sleep, and he kept his body very, very still. He had no intention of bleeding into his body cavity if he could possibly help it. "You go inside your body and have a conversation with that voice inside you," he said later. He suffered some broken ribs and two fractures in his pelvis, and his lungs were bruised and shocked, but, miraculously, they hadn't been punctured. Kevin Hillery may hold a world height record for survival in a fall from a tree. "I'm a member of an exclusive club you don't want to belong to," he said to me. He was released from the hospital four days after the fall.

SILLETT, WALLOWER, SHAFFER, AND SEVERAL OTHER FRIENDS STAYED at the hospital for part of that night, and then returned to Wallower's house. At sunrise, after they'd gotten reports that Hillery was expected to live, they went to the spot where he'd fallen. Frost had formed overnight, and an outline of Hillery's body had appeared as a shadow in the white ground.

The ground was composed of a deep layer of spongy material called duff. Duff is a mixture of decaying needles, twigs, tiny roots, fungus, and other vegetation. Most of it had fallen from the forest canopy. It was a natural mattress, soft and slightly bouncy, and it had saved Hillery's life. "I love duff," Hillery said to me. What had also saved his life was the fact that he had managed to spin his body around and get his head up as he fell. His clawing at the air had worked; he had a superb sense of balance. Punching the ground with his open hand when he landed had also apparently helped to protect his neck. If he had wasted his time screaming, he would have died.

As soon as he was able to climb trees again, which was a few months after his fall, Hillery decided that the first tree he would climb was the one that he had fallen from. He needed to embrace the cir-cumstances of his fall. He found the broken branches that he had an-chored his rope to. He touched the place where they had broken off, not blaming the tree but himself. He had been hurrying, and he had failed to make sure that his rope was anchored over a strong branch before entrusting his life to it and swinging out into space. It was a mistake that only a professional climber would make, for a beginner at tree climbing would have been much more cautious. Only an easy skywalker would have done that, a pro who had gotten too used to moving effortlessly in trees and had momentarily lost his respect for gravity. A year and a half after his fall, Kevin Hillery won three first-place prizes in climbing competitions at an I.S.A. jamboree. He got married and had a child, and started his own tree-care service in Port-land, called Whole Tree Works.

Sillett couldn't get over the accident. In the months afterward, he began to have flashes in which he heard the sound of Hillery's fall—the hissing sound followed by the impact. It was as if Hillery's fall had become locked in time, and was still happening, and would be hap-pening forever. Once Sillett was driving along in his truck and sud-denly he heard a rising hiss, and he flashed Hillery clawing the air, followed by that deadman boom, and whoosh, and then silence, and then a hoarse sound as Hillery drew a breath. He began to feel sick, and he pulled over to the side of the road and threw up.

NINJA ASCENT

NOT LONG AFTER KEVIN HILLERY'S FALL, MICHAEL TAYLOR (who still knew nothing of Sillett or Hillery) decided to hike up to Duckett Bluff, the grassy overlook in Humboldt Redwoods State Park where, thirty years earlier, Paul Zinke had stood while Al Stangenberger raised a weather balloon along the Dyerville Giant. The bluff provides a view of some of the flats in the park, which are filled with tall trees, including the Canfield Grove, parts of a tract called the Rockefeller Forest, and Founders Grove (where the Dyerville Giant now lay in pieces). Taylor had never visited Duckett Bluff. He looked westward, straight up the throat of Bull Creek into the Rockefeller Forest, and immediately he saw an extremely tall redwood. He looked through a sighting instrument and studied its top. It was about a mile away. It seemed to have a double top—two pointy little leaders. The top was explosively tall, rising far above the surrounding canopy, which was incredibly tall in its own right. He could see, at first glance, that this was one of the tallest trees in the forest. It might well be the world's tallest tree. He could feel it in his bones. In silence, in his mind, he named it the Humboldt Tree.

Taylor moved around on the bluff, shooting three compass angles toward the tree. Over the next month, he bushwhacked along his compass lines, following them into the Rockefeller Forest, searching for the base of his Humboldt Tree. The lines took him to the north side of Bull Creek, and he ended up in a flat, where the soil was rich and wet, perfect for the production of redwoods of extreme height. He measured one tree after another, but none of them was tall enough to be the tree he'd seen.

Around the middle of April 1994, Taylor was still searching for the Humboldt Tree. He knew that he was getting close to it. He eventually came to a patch of rising ground on the north side of the creek, where the compass lines seemed to converge. The ground rose into a low hummock, and out of the hummock grew a magnificent redwood. He measured it and got a reading of 371 feet. This was it, the Humboldt Tree.

It reminded him of the Dyerville Giant. It was both an extremely large tree, a titan, and exceptionally tall. It leaned over, too, just the way the Dyerville Giant had done—but it leaned even more than the Dyerville Giant. And it had a weird shape: eighty feet above the ground, the tree divided into a Y. An extra trunk had sprung out of the main trunk and shot straight upward for 260 feet, running next to the main trunk, while the main trunk leaned out over the ground. From Taylor's rough measurements, the tree looked to be perhaps five feet taller than any other known tree. He felt a sense of victory, but even so, he couldn't prove anything. Because the tree leaned over so far, he didn't know how accurate his reading was.

HE KEPT THE DISCOVERY TO HIMSELF FOR SEVERAL MONTHS, BECAUSE he wasn't sure of his measurements and he dreaded making a fool out of himself. Finally, during the summer of 1994, he began quietly telling people that he thought he had found the world's tallest tree. One of the people he talked to was a Portland-based photographer and filmmaker named R. Steve Foster, who was himself a tree climber. Foster happened to know Steve Sillett, and he told Sillett something about Michael Taylor, and said that Michael believed he had discovered the world's tallest tree.

Sillett thought the man had to be a kook. But he called Taylor anyway. "What's this about?" he said. "You can't know it's the tallest tree."

"All right, but you know those claims by National Geographic and other people that they've surveyed the redwoods for the world's tallest tree?" Taylor answered. "It's bullshit. It's complete bullshit. Nobody's ever actually gotten the survey work done."

"What about the Libbey Tree?" Sillett asked—the Tall Tree, that is.

"Its top has died," Taylor said.

"I know." Sillett had stared at the Tall Tree himself. "I've never trusted those claims, either."

"Paul Zinke did some important work," Taylor went on. "I talked with Zinke. He shared some of his data with me. I've been doing the survey work myself with Ron Hildebrant."

"Who's Ron Hildebrant?"

Taylor explained who he was. "All the existing data comes from the 1960s. The data's worthless. The trees have grown. Some of them have fallen, like the Dyerville Giant."

"Okay, what's the deal with the Dyerville Giant? How tall was it, do you think?" Sillett asked. He had also visited the Dyerville Giant right after it fell. His and Taylor's paths had almost crossed there, apparently.

"In my opinion, the Dyerville Giant was the world's tallest tree," Taylor said, and he explained how he had figured it out by looking at the height of scratches on a nearby tree.

Sillett had begun to think that Taylor might be for real. Not strictly a kook, anyway. The only truly accurate way to measure the height of a tree, however, is to stretch a measuring tape along its trunk from top to bottom. Sillett told Taylor that he might have to climb Taylor's tree and measure it himself. Taylor said that sounded like a good idea, and he offered to guide Sillett to the location of the Humboldt Tree.

AT THE TIME, THE GOVERNMENT HAD NO POLICY WITH REGARD TO who might be allowed to climb redwoods. Nobody had ever asked to climb them. As Sillett thought about it, he concluded that if he did ask officials for permission to climb and measure what was possibly the world's tallest tree, the answer might be no. Therefore he plotted out a ninja climb—a stealthy climb, done suddenly and quietly, and partly at night. There would be a team of ninja tree climbers. They would be expert climbers. They would use black-colored ropes—military ropes. They would wear olive drab or other dark clothing. They would conceal their vehicles and equipment. Ever since the climb of Nameless, when he was in college, Sillett had devoted most of his climbing to Douglas-firs. But he had never gotten Nameless out of his mind, and now he was going back to the redwoods.

On Friday, September 16, 1994, the Humboldt Tree team piled two pickup trucks full of ropes and climbing gear and drove from Oregon to California. It was an odd crew. Sillett was the expedition leader, and would be the lead climber (the first one up the tree). As the second climber—his partner for the ascent—he appointed a man named Scott Altenhoff. Altenhoff was twenty-seven at the time, and worked as a climbing grunt for a tree-care company in Corvallis. He also climbed conifers and harvested seed cones from them (the seeds are for growing tree seedlings). The team also included Steve Foster (the filmmaker who had first learned about the tree from Michael Taylor); a photographer named Jack Popowich; and Joe Cordaro, a rock climber who had once gotten involved in a Greenpeace demonstration in which he and seven other climbers suspended themselves from long ropes hanging from the Astoria Bridge, which spans the Columbia River between Washington and Oregon, blocking a U.S. Navy warship from passing through. They held a banner that told the Navy to take its nukes somewhere else. Eight hours later, they finally climbed back up their ropes to the bridge, where it seemed as if every cop in the state was waiting to arrest them.

Amanda LeBrun was asked to join the ninja climbing team, but she had a new job teaching middle school, and reluctantly said that she couldn't go. Kevin Hillery and Douglas Wallower didn't join the team, either. Hillery was still recovering from his fall, and Wallower just wasn't interested. "I'm really a Douglas-fir guy, not a redwood guy," he said.

THE TEAM STOPPED OFF IN GRANTS PASS, IN SOUTHERN OREGON, TO visit New Tribe, Inc., a maker and supplier of recreational tree-climbing gear. New Tribe was founded and then operated by Tom Ness and Sophia Sparks. Tom Ness is credited with inventing many important items of tree-climbing gear that are now in use around the world. He is a laconic man with a seamed face and a thoughtful way of speaking, and he has a dry sense of humor. His father, Sonny Ness, was a commercial arborist, and Sonny taught Tom the arborist climbing technique when Tom was ten years old. Many of Tom Ness's inventions are made of fabric. Sophia Sparks is a specialist in fabrics. She sewed many of Ness's prototypes on an industrial sewing machine, and she contributed ideas to his designs.

In the 1980s, Ness and Sparks were living in San Francisco in a former Sears Roebuck building. Ness was designing and making kitchen tools out of manzanita wood, and he and Sparks were earning a modest living selling the tools out of a booth they set up at street fairs. In 1983, Tom Ness came up with an invention, something he called the Ness Insulated Zipper Cape. It was a soft, thick, padded cape that you could wear over your shoulders while you were hiking. If you got cold, or wanted to stop and rest, you could hunch down under the cape and it would drape down around your body until it touched the ground, forming a sealed cone of warmth. You could get warm while enjoying the view, since your head stuck out of the top of the cape. You could also drop your pants and relieve yourself underneath the cape in privacy and warmth. Tom Ness had invented a portable yurt and outhouse with a view.

There was, however, a problem with the Zipper Cape. You couldn't wear a knapsack on your shoulders while you were wearing it. In response to this challenge, Ness and Sparks invented a knapsack that could be worn underneath a Zipper Cape. They called it the SaddlePack, and it was fully compatible with the Zipper Cape. Unfortunately, the combination of the Zipper Cape and the SaddlePack wasn't really marketable. "When you wore them together, they made you look like a hunchback," Tom Ness explained to me.

They decided to sell the SaddlePack as a stand-alone item, and they founded New Tribe to sell it by mail. The name of the company was meant to pay homage to the indigenous inhabitants of North America, for the Indians had been adept at using local materials to make cool inventions, and this, they felt, was the spirit of New Tribe. The SaddlePack was a blowout success for New Tribe: in two years, the company sold more than a hundred of them.

Ness, who hadn't lost the tree-climbing skills his father had taught him, was visiting an aunt in Ohio one day, and he trimmed some trees for her. He hated the climbing harness that was then in use by arborists, and he came up with what is known as the New Tribe tree-climbing saddle. It's a beefy, comfortable harness especially designed for tree climbers. He sewed the first New Tribe tree-climbing saddle on Sparks's industrial sewing machine.

When he went hiking in Marin County, Tom Ness loved to set up a hammock and sun himself. But he didn't like any of the hammocks he could buy, so he dreamed up a new one, and Sophie Sparks sewed it, adding many practical details of her own. Nowadays, their Treeboat hammock is used for sleeping in treetops. It's made of olive-drab nylon, with straps at each corner that can be attached to the branches of a tree. Climbers who sleep in Treeboats keep themselves tied to the tree while they're asleep, so that if they accidentally roll out of the Treeboat they won't fall to the ground.

When tree climbers want to get a rope up over a branch, they sometimes throw a light, slippery line first—a throw line—and they use the line to pull a rope over the branch. The line has to have a weight on the end so you can throw it, and for many years tree climbers had been attaching a sort of heavy rubber avocado, called a throw ball, to the line. Ness disliked the rubber avocado. So he invented a small, soft bag full of shotgun pellets, a throw bag, and Sparks came up with the best shape of the fabric for it. The throw bag is about the size of a pickle, and is used by most tree climbers worldwide. It has completely supplanted the rubber avocado.

In 1991, Ness and Sparks, a tall, dark-haired woman with blue eyes and a brisk and intelligent manner, moved New Tribe's operations from San Francisco to Grants Pass, where they bought an abandoned grange hall. With painful slowness, they rebuilt the Grange while establishing their business of making and selling tree-climbing gear. By 1992, the New Tribe Grange had become "a construction site occupied by a manufacturing business," as Ness described it. He installed a metalworking shop in the Grange, and he continued to come up with new inventions.

Sparks's practical, businesslike manner seemed to dovetail with Ness's dreamy tinkering. She set up her industrial sewing machine in the main hall of the Grange and began sewing tree-climbing saddles and Treeboats. The manufacturing area was also their living quarters. Ness hung a Treeboat at ground level inside a wooden frame and slept in it. Sparks didn't find the Treeboat comfortable, so she slept nearby, inside their old street-fair booth. The booth contained what Ness calls a flat bed. "I call it a flat bed to distinguish it from a sleeping hammock," he said to me. "We got together quite regularly on the flat bed inside the fair booth." That was until they had a falling out with each other, and they subsequently split up, and Sparks began running New Tribe on her own.

When Sillett and his companions walked into the New Tribe Grange in the fall of 1994, they found Tom Ness in a state of keen interest. He had recently learned about a knot called the Blake's hitch. It had been invented a few years earlier by a California arborist named Jason Blake, and the knot had been sweeping through the tiny world of tree climbing like a virus. The knot slid and grabbed on to a rope really well, it was safe, and it didn't come untied by accident.

Tom Ness demonstrated the Blake's hitch to the ninja team members. They had arrived at the headquarters of New Tribe at a moment when the art of climbing trees and the field of canopy science were taking quantum leaps forward. New technologies and methods of gaining entrance to the canopy were being developed rapidly. It was a time not unlike the early days of scuba diving, when Jacques-Yves Cousteau announced that an unexplored world lay below us in the sea, and that humans could go there. The canopy lay above, waiting unseen.

. . .

THE CLIMBING TEAM BOUGHT TREEBOATS FROM NESS AND SPARKS, and they drove to Humboldt Redwoods State Park and pulled off into a parking lot, where a white Pontiac Le Mans with a smashed rear window was waiting. Michael Taylor stepped out of it.

Steve Sillett wasn't impressed, and thought he looked nerdy. He noted the fact that Taylor had a fancy pocket calculator.

They unloaded the climbing gear from the trucks and piled it out of sight, then they parked the trucks on a dead-end road. Taylor had established a trail to his tree using tiny bits of orange tape stuck to things in the forest. They shouldered their gear and followed Taylor's marks, in growing darkness. They didn't use flashlights.

The Humboldt Tree loomed out of the forest floor. It was a massive and somewhat sinister-looking thing, leaning over starkly. All the other redwoods around it were perfectly straight and vertical. It was like a drunk at a church meeting, and the climbers didn't like the look of it.

"This bastard looks like it's ready to fall at any moment."

"It definitely doesn't look healthy."

"Michael, you've found a doomed tree," Sillett said to Taylor. "It's a death trap."

"Steve, you're such a whiner," Jack Popowich said.

"This tree is solid, I'm telling you."

"It's been here a *long* time. What's the probability it will fall in the next twenty-four hours? Near zero."

"I don't know. This thing looks sketchy. Like it'll just tip over if we climb it."

"There's five of us. What's our combined weight going to be, anyway?"

"It's not going to *fall* under our weight. Our weight is nothing compared to this tree."

Sillett was circling around it, not saying much, trying to get a feel for it both as an organism and as a structure to be climbed. Something flashed, a bit of metal. He bent down and found a metal tag with the number 12 on it, nailed into the tree with a tack. "What's this?" he said.

"Oh, my gosh! This tree is one of the Peas in the Pod!" Taylor exclaimed. He hadn't seen the tag. It was one of the trees that Paul Zinke had told him about—one of the giant redwoods Zinke had discovered in the sixties.

"I think it'll be fine to climb, as long as it's not windy tomorrow," Sillett said.

The team settled down under the tree to camp for the night. They put up their Treeboats, stringing them at ground level from small trees. Michael Taylor had to open the C&V Market the next morning, and he drove back to Eureka and spent the night in his room. The climbers ate handfuls of gorp for dinner.

Steve Sillett lay back in his hammock, staring up the trunk. The moon was rising, and the crown of the tree was ghostly against the clear September sky. It was an ominous presence. He couldn't sleep. He rolled this way and that.

The next morning, at earliest light, Sillett got up and began moving around the campsite. Scott Altenhoff was up, too. It turned out that Altenhoff hadn't slept much, either.

Scott Altenhoff was a tall, thin, introspective man with a wiry frame. He had long hair and a beard, and he was wearing a kind of billowy Renaissance-style beret. He was feeling unready for what lay ahead. He had never climbed a redwood. None of the others had, either, except for Sillett, years earlier.

The Y shape of the Humboldt Tree struck them as very peculiar. The tree had an asymmetrical, forked form, consisting of a main trunk and an extra, secondary trunk shooting out of it. The tree had begun to lean over five or six hundred years earlier. At that point, it had grown a new trunk, and the trunk had gone straight up out of the lower part of the leaning main trunk. Now the extra trunk was very large. It seemed to be acting as a counterweight, keeping the tree from toppling over.

They decided to try to get up the extra trunk first, and then to make their way horizontally through the air to the main trunk somehow, and continue on to the top. Sillett had brought a hunting bow with him. He took it out of its case and began walking around the base of the Humboldt Tree with Altenhoff while they stared up into the crown. The hunting bow had a fishing reel attached to it, loaded with fishing line. They were looking for what is known as a window. A window is an opening in the foliage that provides a passage for an arrow to pass over a strong branch. They finally settled on a window far up on the extra trunk. Sillett tied the fishing line to the end of a blunt-tipped arrow. He drew the bow and fired the arrow, aiming it slightly above the target branch. The branch was two hundred feet up.

The arrow flew upward, trailing the fishing line behind it. It hit something, bounced off, and fell down through the tree to the ground.

Altenhoff asked Sillett if he could try a shot. He fired and missed.

"The knight errant Altenscrotch couldn't hit the wide side of a barn," Sillett joked.

Altenhoff shrugged and handed the bow back to Sillett, who shot several more times, until he finally got the arrow through the window. The arrow passed over the target branch and fell back to earth, pulling the fishing line after it. The fishing line was now draped over the branch.

They tied a nylon cord to the end of the fishing line and pulled the cord over the branch. To one end of this cord they tied a black-colored climbing rope. The rope was superthin and superstrong. It was a type of rope sometimes known as an assault line, or a black tactical rope. The rope is favored by military Special Forces for vertical operations at night. It would be nearly invisible as it hung along the trunk of a redwood.

Sillett and Altenhoff used the cord to haul the black rope up into the tree, over the branch, and back down to the ground. They tied one end of the rope to a small tree that grew nearby. The other end dangled loosely down to the ground from the branch. Sillett's plan was to climb up the loose end of the black rope to the branch using ascenders. The black rope had been rigged as a ground-anchored climbing rope.

While the others watched, Sillett put on a climbing saddle and a helmet. He clamped a pair of Jumar-type ascenders to the black rope and began jugging upward rather slowly, his body spinning on the rope. He moved upward through a middle zone of space among the trunks in the grove. Because the rope was thin and camouflaged, it looked as if Sillett were climbing up on nothing.

JACK POWOWICH WAS FEELING NERVOUS AS HE WATCHED STEVE SILLETT ascend seemingly into empty air. Popowich was a small, tough man with a handlebar mustache who knew some acrobatic tricks. From a standing-still position, he could do a backflip in the air and land on his feet. He enjoyed walking into a bar where he wasn't known, getting himself stinking drunk, and then, in a loud voice, claiming that he could do a backflip and land on his feet. Pretty soon he would start taking bets. He won every time.

This damned tree looks like it's getting ready to fall, Popowich said to himself. The climb was surely going to make Steve Sillett famous, but it was scary as hell to watch him do it.

"I don't know about this tree," Sillett's voice came floating from above. He sounded exasperated.

"Quit whining," Popowich shouted up, cupping his hands around his mouth. "What's the biggest whiner in the woods crying about now?"

Sillett tried to ignore him. Eventually, he got up two hundred feet, and he reached the anchor branch, over which the climbing rope was draped. He could go no farther on the black rope.

Sillett looked around. The tree seemed healthy here. The branches were alive, covered with foliage, and strong-looking. "This tree looks great," he shouted down. He was carrying a short length of arborist rope with him. He tied a loop of the rope around a branch (for safety), detached himself from the black rope, and sat on the branch. He shouted down to Altenhoff that the main rope was free and he could start climbing.

Altenhoff put his ascenders on the black rope. As he got up off the ground, he began spinning around. The act of vertical entry into the redwood canopy was literally a dizzying experience for him. When he got to the top of the black rope, he detached from it and sat on the branch next to Sillett. Sillett took up the end of the black rope and yelled down to the ground, telling the team to untie the rope from the ground. He would drag it with him, because they would need to use it again higher up. Then he started free-climbing upward along the secondary trunk, moving from branch to branch. Altenhoff followed Sillett up the trunk.

Scott Altenhoff had majored in classics in college, where he had mastered ancient Greek and Latin, and he had fallen in love with the works of Homer and Aristotle. At one time he had planned to become a professor of classics. His parents couldn't afford the cost of college, and he worked his way through school by getting a summer job as a climber in a tree crew. Eventually, he decided that he wanted to have a profession in which he could use both his body and his mind, and he became a certified arborist.

As Scott Altenhoff entered redwood space for the first time in his life, Aristotle's *Metaphysics* came into his mind. Aristotle believed that cause and effect happen in nature because objects tend toward realizing their potentialities. For example, according to Aristotle, an apple falls from a tree because it is in the nature of objects to move toward the earth. The apple achieves its potentiality by falling to the ground. (Isaac Newton rejected Aristotle's *Metaphysics* and came up with physics instead: the laws of gravity, the mathematical equations that describe gravity's action between all objects that have mass.)

Altenhoff thought that this tree had Aristotelian drive. It had been striving possibly since the time of Aristotle to become what it was. It was a living thing in a state of flux, forever becoming what it must become. Trees are unconscious, self-directed life forms, driven into their elaborate shapes by the programming in their genetic code, and by their responses to sunlight, wind, water, accidents, fire, insects, and disease. The Humboldt Tree seemed overwhelmingly alive as it crowded upward into the air, seeking to drive its form into space and move toward the light, to cast its shadow over lesser trees and take their light from them, and to throw its seeds into the world, and so make itself immortal.

A word from Greek tragedy kept running through his mind: ὕβρις. Hubris. The violent and unnecessary human pride that tempts the gods to crush a man. This was no place for humans, and he could feel it. He was just a little thing crawling on the tree. He was a flea on Agamemnon.

■ ■ ■

SILLETT AND ALTENHOFF WERE CLIMBING WITHIN AN ARM'S REACH of each other. They climbed from pitch to pitch, until the secondary trunk came to an end at 260 feet above the ground. They clung to it, looking over at the main trunk and considering their next move. Sillett tossed a throw bag, which was trailing a throw line, into the main crown of the tree. It fell over a branch there, and he let it run to the ground. He tied the black rope to the throw line, and someone on the ground pulled the line, which dragged the black rope into the main crown of the Humboldt Tree and the end of the rope down to the ground, where the team members on the ground re-anchored it. When that was accomplished, Sillett and Altenhoff used their short ropes and the arrangement of the long black rope to skywalk across to the main trunk through the air. Once they had gotten themselves safely over to the main trunk, they free-climbed up to the top of the tree, keeping slack ropes tied to branches as they went, to save their lives in a fall.

Sixty feet below the top, the main trunk divided into two thin, pointy leaders growing side by side. Sillett climbed up the taller of the two leaders and emerged through the top surface of the redwood canopy. Altenhoff stayed about ten feet below him. The Humboldt Tree was only three inches in diameter near its top, where a burst of foliage trembled softly in a sea of moving air. The tree was vigorous, green, and healthy, and it looked as if it could live for another thousand years. Explosive crowns of other supertall redwoods bubbled across the view, at eye level. Above the canopy stretched the milky blue of a California sky.

"Scott, I'm standing at a place where my eyes are above the topmost foliage of this tree," Sillett said. Above the world's tallest living thing.

Oooh, oooh, Altenhoff thought. He's scampering around up there. He hoped that Sillett wasn't damaging the tree's leader.

Sillett took a tape-measuring line from a bag and attached a weighted bag—a throw bag—to it. He reached upward and caressed the topmost sprig of the tree lightly with his fingertips. Then he let the weight go, and the tape line slithered down the trunk of the tree. Altenhoff descended with the tape, and they measured the vertical height of the tree, from the tip of its highest foliage to the ground.

It was 359 feet tall—not quite the tallest tree.

"Michael Taylor blew it!" Sillett shouted down. The others laughed, and some of them felt really sorry for Taylor. They rigged a rope running to the top, and one by one they all climbed the tree. Steve Foster carried a sketch pad and a pencil with him, and he spent the afternoon exploring the Humboldt Tree and making drawings of its branches and trunks.

THAT AFTERNOON, MICHAEL TAYLOR SHOWED UP AT THE BASE OF THE Humboldt Tree, anticipating a triumphant welcome as the discoverer of the world's tallest tree. Instead, Sillett chewed him out. "Dude, you got confused by the lean," he said.

Taylor apologized. He had brought a couple of foil bags with him, and he handed them to the climbers. The bags contained roasted chickens from the supermarket.

The climbers thanked Taylor and began tearing into the chickens. They had had nothing to eat all day except for a few handfuls of peanuts and raisins. Sillett dug in, and his perception of Taylor got a little warmer. What kind of person brings you a roasted chicken as a gift? Michael Taylor couldn't really afford this. He suddenly began to like him. "We've got to get you up into this tree," he said to him.

"It's just not right that you haven't climbed it yet, when you dis-covered it," Scott Altenhoff added. "It's a shame that you've never ex-perienced the redwood canopy. We've got to remedy that."

The climbers started looking over their gear so that Michael Tay-lor could get harnessed up.

"That sounds good," Taylor answered. He was looking at the ground, and his face had turned the color of suet. The climbers saw it: he was terrified of heights. "I don't think I could do it for physical or psychological reasons," Taylor finally mumbled.

The climbers loaded their packs with Treeboats and sleeping bags, and ascended the main rope one at a time, hauling their gear with them. Taylor stayed at the base of the Humboldt Tree and watched as they receded into redwood space, until they were lost to his view. He felt a sense of longing and loneliness as they left the ground. He wanted to go with them, but it was something he couldn't do. He got back into his car and drove home. It was Saturday night, September 17, 1994.

THE CLIMBERS TOOK OFF THEIR BOOTS AND MOVED AROUND BARE-foot in the top of the tree, and some of them strung up their hammocks as high as possible, between the two leaders. Joe Cordaro placed his hammock twenty feet below the top—thirty-four stories up. Steve Sillett placed his hammock just below Cordaro's. Each climber wore his harness and kept himself tied to the tree with a safety rope. There was a lot of joking and horsing around.

In September, the weather along the North Coast is usually warm and dry. The moon rose before sunset, round and full, and it glowed in the air. A breeze stirred the leaders and they moved back and forth, rocking the hammocks.

Sillett lay in his hammock, looking at the moon. He didn't like Michael Taylor's name for the tree, and he thought that a better name for it would be Telperion, or the Tree of the Moon. It was a name taken from *The Silmarillion,* Tolkien's mythic history of Middle-earth. Sillett mentioned his idea to the others. They thought it was rather like Steve to give the tree a weird name. They settled into their hammocks, which hung far out in space over the forest floor, because the tree leaned so much. Telperion resembled a parked construction crane. A faint breeze played with the tree, and it rocked slowly. They fell alseep to its movement.

DETONATION ZONE

SOMETIME PAST MIDNIGHT, STEVE SILLETT WOKE UP WITH THE moon shining in his eyes. It was so bright that he couldn't sleep. It was a calm, clear night, with the stirrings of a faint breeze. He sat up in his Treeboat, looking around. He wanted to climb to the top of the tree, to see the tallest forest canopy by moonlight. It was almost certain that no one in history had ever seen it from the top and from within by the light of the moon.

He stood up in his hammock. He was barefoot, but he was wearing his climbing harness and was attached to Telperion with a rope. He unclipped the rope, detaching himself from the tree. He stepped out onto a branch and free-climbed up to Cordaro's hammock. He stood up on the lip of Cordaro's hammock, reaching for another branch.

Cordaro let out a high-pitched squeal. "What are you doing, Steve?"

"I'm just trying to get past you."

The others began laughing.

He laddered his way barefoot to the very top of Telperion. He didn't use a rope, and he felt that any hominid with any dexterity could have pulled off the climb. When he got to the highest solid branch, he stood on it and looked around at the other giants in the grove. Only their tops were visible, forming rocket plumes and billows. A sad feeling came over him. He touched the bark of Telperion with his fingertips, and thought how the tree would live so much longer than he would.

Folds of the California Coast Ranges lay to the west, and beyond them the sea. The breeze became fresher and the air began to feel cold. After a while, Sillett climbed down and got into his sleeping bag and went to sleep.

Soon afterward, the temperature dropped abruptly. Clouds began to pop up over the ridges and began spilling down into the redwood valleys, pouring through Panther Gap, to the west. The wind increased to a stiff breeze, strong enough to make the branches of Telperion bend and wave. A sheet of clouds raced under the moon and obscured its light, but there was still enough light to see the trees. In less than five minutes, the wind picked up until it seemed to be heading to gale strength. Then a misty rain began to fall, making noise and wetting the climbers' sleeping bags.

They sat up in their hammocks, grumbling, but the complaints quickly stopped. They began feeling out the weather, trying to get a sense of what was happening. A sharp squall was coming in from the ocean, and squalls can come in fast on the North Coast. The twin leaders of Telperion began to rock, moving independently of each other, and the hammocks strung between them swayed and twisted.

The climbers wondered how bad this was going to get. Someone commented that it was pretty early in the season for an autumn storm, and someone else wanted to know if anyone had heard a weather report. They were frightened, though none of them wanted to admit it. The wind began coming in bursts, and the rain fell at a slant.

Telperion's swaying became more pronounced. The entire tree was rocking now. The wind increased. Within a few minutes it became a whole gale, a wall of wind coming in from the sea. The rain fell horizontally, a light rain, not heavy, but it felt like pinpricks on their faces. As suddenly as it had begun, the wind slowed down, and the rain fell vertically again. The climbers began to relax, but it was a false pause. There was a roaring sound, and the rain came in sideways as the wind ran up to a gale again.

Though it was dark under the clouds, the climbers could see shapes around them. The redwoods were moving, swaying in different directions and in different intervals of time. Each redwood had its own resonance in the wind. Scott Altenhoff could see the upper trunks of some of the largest redwoods changing from bright to dark as they moved back and forth through the dim shadows of neighboring trees. The upper shaft of Telperion began sweeping through longer arcs, pulling about fifteen feet with each swing, enough to make them feel sick. Its motion was like the tip of a fly fisherman's rod. The movements of the top of Telperion became circular and twisty, or C-shaped—the top of the tree began moving in a cyclical pattern. The weight of the people in the hammocks exaggerated the cycles of the leaders, magnifying the arcs that the tree traced as it floated within the upper surface of the grove.

The climbers were tree surfing in their hammocks.

Telperion began vibrating, too, with a sort of deep subsonic shake. None of them had ever felt anything like this in a tree.

They listened over the wind, but couldn't hear the tree, exactly. The tree was making sounds that were below the range of human hearing. They put their hands on the trunk, trying to feel the vibration, trying to get a sense of their situation. When Telperion rocked away from the direction of its lean, it went into an upward-tending sway, graceful and rubbery, pulling gently against its root system and straightening up a little bit. It would reach the top of its arc, and then it would begin to tip downward in the opposite direction, descending into its lean. When the tree rocked into its lean, the angle of the lean became steeper. The tree went over, leaning more and more, and then it stopped with a lurch. Each time it stopped, they felt something that they couldn't hear, something below their ability to hear it, but they could feel it in their hammocks. Then the tree would start moving in the other direction, straightening up, and there would be a pause before it descended downward into its lean again. Now they began to hear deep sounds—*poom, poom, aloom*—and again the tree stopped, with the hammocks hanging out in space.

"I don't think this tree is doing well," Sillett shouted over the wind.

The vibration or shock every time the tree stopped seemed to be coming from somewhere in the ground at the base of the tree. It seemed to be coming from down in the soil, or from the tree's roots. It felt as if something at the base of the tree was breaking apart.

STEVE SILLETT COULDN'T GET THE DYERVILLE GIANT OUT OF HIS mind: that pancake of roots tipped up into the air, that crater forty feet across. He was also conscious of the fact there were very few standing dead redwoods anywhere in the groves. No rotting skeletons of redwoods standing upright. The floor of the redwood forest was a maze of fallen trunks. Now, in Telperion, the meaning of it became very clear: redwoods fall while they're still alive.

Telperion was a tower of wood thirty-six stories tall, and it was balanced at an angle in moist soil, like a pencil standing in mud. The tree was a lever as tall as an office building, and it was being held upright by a mat of fine roots just two feet thick. Telperion would be exerting unbelievable forces on its root system, he thought. It was not comforting.

They began debating whether they should try to pull an abort and get everyone the hell down to the ground as fast as possible. They considered the difficulty of performing an emergency descent at night during a storm. There were five people in the tree. Each person would have to descend along the trunk in darkness, wind, and rain. They would have to rappel down along the main rope one person at a time. For safety, only one person should rappel on a rope at any time, so it would take a while to get all members of the team to the ground. Someone could make a mistake and fall during the process. They didn't have any communication radios with them—they would have to shout between the top of the tree and the ground, and with the noise of the wind they could get their signals mixed up, increasing the chances of an accident. They certainly could not take down the hammocks. Should they abandon the gear in the tree and stand around on the ground to wait out the storm?

Sillett was the lead climber, and the others looked to him for a decision. As the lead climber, he would be the last person to exit the tree. He considered the fact that two of the team members—Cordaro and Popowich—were rock climbers who had no experience climbing trees. Furthermore, in a wind it's safer in the top of a tree than it is on the ground, because you can't be hit by a falling branch.

They talked it over, and decided that the safest thing would be to stay in their hammocks. The tree was at least a thousand years old and maybe quite a lot older. It had survived storms far worse than this one.

"Don't worry. This tree will live a lot longer than we will. It will be here for centuries," Sillett shouted over the wind.

The clouds thickened over the moon, and the night grew darker.

Scott Altenhoff huddled in his hammock, feeling strangely cozy as he was rocked around: maybe it was just that a sleeping bag gives you a primitive feeling of security, even if the security is an illusion. The trunks were moving independently of one another, working and waving separately. There was no steady point of reference, no horizon, no moon visible. He felt as if he were in the middle of a heaving sea.

Sillett began to consider his safety rope. Tying yourself to a redwood that might fall seemed fairly stupid, he thought. The rope would pull you down along with the tree. But if he wasn't tied to the tree, then he could bail out of his hammock and dive away from the tree if he felt it beginning to tip over. If it started to fall, how much time would he have to unclip and bail out? A few seconds? He would have to leap out of Telperion, kicking away from his hammock while everything was going into free fall. He would have to jump really hard to keep from being sucked down by the passing branches—like the pull of a sinking ship. Then, if he managed to get clear of the falling Telperion, he might plunge down into some dense part of the canopy and be able to grab something as he passed through. It wasn't realistic to think that he could survive by swan-diving thirty-four stories out of a redwood into pitch darkness, yet it would be the only chance of survival if Telperion fell: he knew it would be certain death to ride one of these monsters into its grave. He unclipped himself and prepared to jump. He didn't say anything to the others.

The gale died down as suddenly as it had begun, and the rain stopped.

WHEN THE SUN ROSE, TELPERION STILL STOOD, LEANING INTO THE grove. The climbers rappelled out of the tree as soon as there was enough light to see clearly. While he was descending, Scott Altenhoff caught his long hair in his rappelling device, and it took him a while to free himself. Meanwhile, he dangled and spun on the rope. This occasioned a good deal of laughter from the others, and it also illustrated how tricky a nighttime descent might have proved to be. Once the climbers arrived on the ground, they started joking about being trapped in a leaning redwood in a storm, and they began to doubt that there had ever been any danger.

A month or two later, Sillett attended a scientific conference, where he presented a paper on the climb of Telperion, and he announced that the redwood-forest canopy was now, for the first time, open for exploration. While he was at the conference, another storm passed over the North Coast. When he returned from the conference, he found a message on his answering machine: "It's Michael Taylor. We've got to talk."

He returned the call.

"Steve, Telperion is on the ground."

SILLETT DROVE DOWN TO EUREKA AND MET MICHAEL TAYLOR AT THE C&V Market, and they hiked into the Rockefeller Forest to see what had happened. The fall of Telperion had created a swath of devastation in the forest. Telperion was nearly as large as the Dyerville Giant. It had smashed a smaller redwood to pieces when it fell, creating a debris field that extended in all directions. The root mass of Telperion extended about thirty feet into the air. Its prone trunk was sixteen feet in diameter—almost three times the height of their heads as they looked up at it. Shattered branches and small exploded trees and

LAST DAYS. Telperion drawn by R. Steve Foster soon after he climbed it with Sillett's team. This is the only image of Telperion that exists. Foster based his drawing on measurements and sketches he made during the climb. This is an ideal view. No one would have seen Telperion this way, since the tree was virtually invisible where it stood, with its crown buried in the canopy and surrounded by other redwoods.

great chunks and splinters of redwood had been flung around the hulk of the tree. Blobs of soil ranging in size from baseballs to basketballs had been thrown up to thirty yards when Telperion smacked into the ground.

"The mud splash when Telperion hit the ground must have been simply awesome," Sillett said to me. "We could see the splash mark way up on the trunks of the trees all around." The trunks surrounding the detonation zone were coated with soil sixty feet above the ground, like a bathtub ring.

They walked along the trunk of Telperion, which was considerably longer than a football field. Sillett inspected lichens and mosses on the trunk and limbs, some of the rare lichens of the forest canopy. On the ground, they would die. With the fall of Telperion, a world had passed away.

Taylor could see every inch of the tree when it was on the ground. It made him incredibly nervous, and it gave him a haunting sensation of his own unknown future, his own death. He touched one of the branches where Sillett had anchored his rope, and he imagined himself tying a rope around the branch when the tree was upright, and climbing the tree himself, the way Sillett had done. You could die doing this, he thought. It gave him an eerie feeling that one day he would climb into the trees.

Eventually, they reached the aerial campsite between the twin spires of the main top. They looked for the branches where the highest Treeboats had been hung—Sillett's and Cordaro's. Those branches were nowhere to be seen; they had been driven eight feet into the ground.

"If we had been in Telperion when it fell, we would have become human body parts mixed with the soil," Sillett said. "They would have had to pull our remains out of the ground by yarding up our safety ropes."

Time in the redwoods moves slowly, except when it moves very fast. The humans had outlived Telperion.

3
THE
OPENING
OF THE
LABYRINTH

THE THREADS

AROUND 1980, AN ARBORIST IN ATLANTA NAMED PETER JENK-
ins began teaching all sorts of people, including children,
how to climb trees safely using the arborist climbing technique. Jenkins founded a tree-climbing organization, with a climbing
school, called Tree Climbers International. The classroom of Tree
Climbers International consists of two white oak trees on a plot of
land near downtown Atlanta. Recreational tree climbing—climbing
trees for fun, using the arborist climbing technique—is an evolving
sport, or emerging oddity, that is growing in popularity.

I came across the Atlanta tree-climbing school while I was surfing
the Internet. I had never thought about climbing trees with ropes, and
it seemed weirdly appealing. I got a cheap flight to Atlanta and began
to learn the art of movement in a forest canopy. At the time, I had
never heard of Steve Sillett or of redwood climbing. The threads of
chance and desire that join us into a world of human experience had
yet to ensnare me in the redwoods.

One day in April, I was aloft in the crown of one of Jenkins's oak
trees in Atlanta, seventy feet in the air and hanging from branches,

and skywalking. I was wearing a helmet and a tree-climbing saddle. An instructor named Tim Kovar was hanging from a nearby branch, watching me. It was a cool, blue day, and the wind was blowing, and we were swinging gently on our ropes. In the distance, the rectilinear towers of Atlanta glittered in the sun.

I was carrying a Ness throw bag attached to a length of long, slippery yellow throw line.

"Throwing," I said, and tossed the throw bag, aiming just above a target branch.

The bag flew through the air, pulling the line behind it. It bounced off the branch, went astray, and fell down somewhere out of sight, dragging the line after it.

"Headache!" I shouted.

I gathered up the throw line, threw the bag, and missed again.

Then someone on the ground yelled, "Throwing!" and a throw bag soared over a branch near us, trailing a throw line. The line floated over the branch and settled down through the tree. It was a dead accurate shot, and whoever had made it was a pro. Next, a rope came slithering up and over the branch and went down to the ground—the person on the ground had used the throw line to set the rope in the tree. Not long afterward, Peter Jenkins, the founder of Tree Climbers International, appeared, climbing up the rope using mechanical rope ascenders.

Jenkins is a lean man in his fifties. He has reddish-blond hair, large forearms, and broad, crisp-looking hands. He hung on his rope and touched his feet lightly against the tree, and watched me while I practiced skywalking.

"Not too bad," he said after a while.

"How long is it going to take me to get smooth at this?"

"You can get reasonably good at tree climbing in six months if you climb regularly," Jenkins answered in a thoughtful way. "Each move has to be done on a nearly instinctual basis. It's called muscle memory—the hands and body instinctively know what to do before the mind does. You have to keep a hundred percent focus on what you're doing when you're in a tree. If you lose focus just once, and make one small mistake . . ." His voice trailed off.

"The ground rushes up surprisingly fast," Kovar said.

WHIPPER

ARIE ANTOINE WAS EIGHT WHEN HER MOTHER, ELIZAbeth, died. "Afterward, I remember being so conscious of how sad my dad was," she said to me. "I wanted to make his life as easy as possible." It didn't always work out that way. When Marie was fourteen, she began hanging out with a group of party girls in Kenora, the town where the family lived when it was too cold to stay at the house on the island in Lake of the Woods. It was a girl scene—no boys around. They were all students at the Beaver Brae Secondary School, the town's public high school. Marie and her friends became fond of alcohol, and they would drink anything they could get their hands on from their parents' liquor cabinets— anything that could be replaced in the bottle with water.

One day there was a dance in the middle of the day at Beaver Brae, and Marie and her friends began drinking vodka. Some of Marie's teachers noticed that she was drunk and they telephoned Ronald. Since he didn't have a car, he called a taxi and took Marie home. She went into the bathroom and threw up. She was suspended from school and began to get a reputation around town as a troublemaker.

Ronald Antoine was almost a teetotaler, and he had never been drunk in his life. He didn't punish Marie, though; in fact, he hardly said a word to her about the incident. He encouraged her to practice the piano after school, and to read. He was always up early in the morning, long before Marie, and she would come down to the break-fast table to find her vitamins laid out next to a glass of orange juice. Soon after she was suspended for drinking, she found a letter on the breakfast table as well. In carefully framed paragraphs worthy of a for-mer high-school teacher, Ronald spelled out to Marie why consuming alcohol was an irrational activity.

Marie had a lot of energy, and her father left her another break-fast letter, encouraging her to take up sports and to pursue the piano with more dedication. She joined the high-school track team, and she began giving piano recitals. She became a track champion at Beaver Brae, doing especially well in the two-hundred-meter and four-hundred-meter races. She turned out to be the best runner from Kenora in her class and events. She toured all over Canada, represent-ing Kenora in national track meets, where she got slaughtered. She found out, she said, that a small town like Kenora was no match for Calgary or Montreal.

Her girlfriends stuck together in a pack. She liked their company, but more and more she found herself hanging out with boys. She was thin, and tall for her age, and she loomed over some of them. The boys had crushes on some of the girls, and they enlisted Marie's help in getting messages to them. She also began helping the boys set up dates with girls they liked. She got a terrible crush on one of the boys, but nothing came of it. The boys never thought of her as a date.

There wasn't much to do around Kenora in the winter, so Marie and the boys would throw snowshoes into a car and drive across the ice to an island or somewhere around the edge of the lake, put on the snowshoes, and hike to a more remote spot, where, often enough, there was a cliff. (There are many cliffs around Lake of the Woods.) They would build a fire under the cliff, out of the wind, and make hot cocoa, fool around, talk. One of their haunts was an open area below White Dog Cliff, where there was a great view. The cliff is a hundred feet high and is plastered with ice during the winter.

If things got slow, Marie Antoine would challenge the boys to climb the cliff: "So are you guys wimps or what?" She wore snow-mobile boots—chunky rubber boots—and she would scale up White Dog Cliff and then climb down again. The boys watched her climb, and then followed her. She was at least as good at climbing as any of them. One day, one of the boys fell down a cliff, but he only cracked a rib. Some of the best cliffs for winter climbing were road cuts around various local highways. People driving by would notice a group of teenage boys, and one girl, standing around near a road cut and doing nothing. They were waiting for the car to pass so that they could climb without being caught.

Marie Antoine also began to hang out with a somewhat rougher set of boys. They liked to take their cars out on the ice and do stunts. They would get a car going really fast across the ice, and then pull the emergency brake. The car would go into a spin, doing multiple three-sixties, while everybody inside shrieked and laughed. This was especially fun if the car whomped into a snowbank at the end of the spin. After she got her driver's license, Marie Antoine tried spinning cars a few times. The boys' cars got to be a little trashy-looking from the spinning.

Kenora was small enough so that the town was able to worry, as a town, about what Ronald Antoine's daughter was doing. One day a friend said to him, "I was driving along, and I thought I saw your daughter climbing a hundred feet up a cliff by the road."

Ronald asked Marie about it.

"That must have been some other girl," she said.

He wondered what other girl in Kenora was capable of climbing a hundred-foot cliff in the winter. One morning she came down to find yet another letter by her vitamins. It was long, logical, and very detailed. In it, Ronald explained why climbing ice-covered cliffs is irrational.

"This was excruciating for me," she said.

MARIE ANTOINE APPLIED TO THE UNIVERSITY OF OREGON, IN EU-gene, because it was known to have a great track team, and she was accepted. When she arrived she found that she wasn't as tall as she had thought she was, relatively speaking. The university had been re-cruiting female track stars from top high-school track teams all over the United States. "I couldn't compete against these giant American women," Marie said.

She thought that she might become a poet or a writer, and she decided to major in English. After writing some poetry and taking quite a few English courses, she changed her mind and became interested in nutrition. She transferred to Oregon State University, in Corvallis, which had a nutrition program. She got into rock climbing as a sport there, and she climbed mostly with men. She began to do lead-climbing—climbing as the leader, advancing up a rock face above a climbing partner. One day Marie Antoine was lead-climbing up a cliff in the Cascade Mountains, with a male belaying partner standing below her on the ground. He got distracted—he was looking at some woman, she thought—and, without either of them realizing it, he allowed a lot of slack rope to accumulate, rather than keeping the rope fairly tight, as he should have. When Marie was almost fifty feet above the ground, right at the redline, she unexpectedly lost her grip and fell. She went into free fall. Her partner began frantically hauling in the rope to try to stop her fall, but it was too slack. For Marie Antoine, time seemed to stretch; she saw the cliff rushing past, and she could feel the ground coming up to break every bone in her body. She thought that she was going to die. At the last possible instant, just before she hit the ground, her partner managed to stop her fall, and she bounced upward on the rope as if she were on a bungee cord. She had taken a screaming forty-foot whipper that had nearly ended in her death.

The whipper snuffed out her courage for advancing up a rock face as the lead climber. It also gave her a fear of hanging on a rope high above the ground, and she lost her faith in climbing partners.

While Marie Antoine had been rock climbing, she had often noticed trees growing along the cliffs. She had been able to look horizontally into their crowns, and had been close enough to some of them to almost reach out and touch them. She would often pause in climbing, staring into the trees. "They were Douglas-firs dripping with lichens and mosses and things growing on them," she said to me. She began dating a fellow student who will be named in this book Ted Eldon, who enjoyed backpacking in old-growth forests, but he wasn't a climber. They backpacked together in the Oregon woods, admiring the trees.

At the start of her junior year at Oregon State, Marie Antoine signed up for a number of classes in nutrition. Then, the night before she had to declare her major, something happened to her. She woke up that morning knowing that she was going to be a botanist. She was going to study plants, especially trees. Marie Antoine has never been able to explain, completely, what it was about trees and their crowns that so fascinated her. It had something to do with the lure of secret places high above the ground, places you have to climb to.

As a college student taking botany courses, she heard lectures on lichens, and she learned about Bill Denison's discoveries in the canopy regarding Lobaria, or lungwort—how it fertilizes old-growth Douglas-fir forests. By then, Bill Denison had retired from the faculty of Oregon State, but he was a well-known figure, famous to botany students. Marie Antoine became entranced with Lobaria and fascinated with Denison's discoveries. She decided that she would study lichen, especially Lobaria. She had been caught by a thread of intellectual fascination with lungwort.

CATHEDRALS IN THE RAIN

AMANDA LeBRUN BEGAN TO REALIZE THAT THE PASSION THAT her husband, Steve, had for climbing tall trees was getting in the way of many of the basic elements of human life: love, personal relationships, their marriage, the possibility of having children someday. She felt that Kevin Hillery's fall had been a warning, and that the irresponsible, macho atmosphere of climbing that Steve had gotten himself into could lead to her becoming a widow.

Sillett was away from the house much of the time, climbing along the Middle Santiam River to collect and study lichens, or working on his dissertation at the Oregon State campus, in Corvallis. He had begun to work with a young woman who acted as a climbing assistant, since LeBrun was too busy with her teaching to climb. Sillett and his assistant climber went camping for weeks together alone in the forest along the Santiam River. They were climbing trees together every day. One day, after a long climb, they took off their clothes and went swimming in the river, and they ended up inside the tent together. Then she stayed in the house with Sillett and LeBrun.

LeBrun began to suspect something. She confronted both Sillett and his assistant, accused them of having an affair, and demanded that it end immediately. He said they had only been fooling around, that it wasn't serious. Sillett was twenty-six, and LeBrun had been his first steady girlfriend and college sweetheart. He hadn't had that many sexual experiences with other people, and he was vague about what he had actually done in the tent with the woman. "It was just physical," he said. He felt that what had happened was the result of two people of the opposite sex camping together. "It was a few incidents of poor judgment out in the field." Amanda LeBrun was devastated. They had been married for little more than a year.

AFTER THE CLIMB AND THEN THE FALL OF TELPERION, STEVE SILLETT asked Michael Taylor if there were any other redwoods that Taylor knew about that might be interesting. Taylor said he'd be happy to show Sillett what he had discovered, and Sillett drove down to Eureka again, and he and Taylor met at the C&V Market. Taylor took him on a grand tour of the groves and trees that he had discovered. He showed him trees that no one else knew about except for Ron Hildebrant. He bushwhacked with him all over Humboldt Redwoods State Park, and showed him Laurelin, Alice Rhodes, Paradox, and many more. Taylor had named the tree Paradox because it had a paradoxically thin trunk—it didn't look like anything special from the ground, just another redwood, but it had the proportions of a needle, and it stabbed upward and then burst into a spray above the canopy. Sillett had never seen anything quite like Paradox. Taylor took him miles into the forest and showed him a gargantuan, stovepipe-shaped redwood with a mushroom top and a cave inside its roots. The tree is called Gaia. Sillett thought that Gaia might be older than the Terex Titan (the half-rotten hulk that Taylor had discovered at Prairie Creek). Gaia might be more than three thousand years old; it might date from the time of the invention of cuneiform writing in the Middle East, but who could say?

Sillett realized that Taylor came close to having a photographic memory for the structure of a forest; he carried a map inside his brain of every place he had ever gone in the redwoods. Taylor, Sillett noticed, also had a passion for accuracy. His measurements were surprisingly careful and precise; they were reliable, and Sillett was impressed with Taylor's intellectual honesty—Taylor never exaggerated things. Taylor's training in engineering and forestry, as well as his love of gadgets, had put him in a good position to explore the redwood forest.

At Prairie Creek Redwoods State Park, Taylor led Sillett into the Atlas Grove. There were other redwood titans near the Atlas Tree that Taylor had discovered and given oddball names. What Taylor had found, Sillett realized, was a wonderland of nature, an unexplored Grand Canyon of botany. He thought that the Atlas Grove would be a good place to begin his life's quest of exploring the world's tallest forests.

Steve Sillett received his Ph.D. in botany in 1995, and, with Amanda LeBrun's encouragement, he applied for a teaching job at Humboldt State University, in Arcata, where Taylor had been studying on and off. He longed to be near the redwood forest canopy, but he didn't think he had a chance of getting the job, and he was very surprised when he did. In January 1996, he towed a used house trailer down to Arcata and parked it at a campground, and lived in it while he took up his faculty duties. LeBrun had to remain in Corvallis until the end of the teaching year, in June.

Arcata is a rainy, remote town, ninety miles south of the Oregon border, with narrow houses under pewter skies. Forested ridges rise around the town, spiky with redwoods and often wrapped in clouds. In the town square, young people wander about, wearing backpacks, and they have walking sticks and dogs. In the Arcata Co-Op, around the corner from the town square, you can buy herbal cures for anything from malaise to melanoma. Humboldt State University sprawls up a hillside above the town. It has 7,500 students, the majority of whom are undergraduates, although it offers master's programs in various areas. The university does not award Ph.D.s.

Sillett began teaching courses in general botany—lecture classes with a hundred students in them. He found that his teaching duties ate up more time than he had expected. He had never taught before, and the work of developing courses, counseling students, grading exams, attending committee meetings, and dealing with colleagues in the department kept him from doing scientific research. All through the spring of 1996, he didn't climb in the redwoods. He worked in his office, often putting in seven days a week, while he was living in the trailer in the campground. In place of research, the best he could do was to go on occasional hiking trips with Michael Taylor into the redwood forests, gazing up the trunks and wanting to go there. He applied to government authorities to climb the redwoods.

The National Park Service, and the California State Parks Department, partly in response to Sillett's request to be allowed to climb redwoods, established an annual climbing season and a system of permits, for scientists to work in the trees. The redwood climbing season began on September 16 and ended on the last day of January. It was only four and a half months long, running through the fall and early winter. The rules were put in place to protect the northern spotted owl, and an endangered seabird called the marbled murrelet, which lays its eggs during the spring and summer on the branches at the tops of old conifers, including redwoods. Thus, that spring and summer, Sillett couldn't climb the redwoods even if he had had the time. He was pretty sure that nobody had entered the redwood canopy since that night in September 1994 when he and his ninja friends had ridden out the storm in Telperion.

Sillett's first task in trying to understand the redwood canopy would be to describe the things that live there. Putting together a basic picture of an ecosystem or habitat and what lives in it is called descriptive natural history. Descriptive natural history is something that the great explorers of nature did—John James Audubon did it when he traveled through North America collecting and drawing birds, and Charles Darwin did it as a young man sailing on the H.M.S. *Beagle* to, among other places, the Galapagos Islands.

IN SEPTEMBER, AT THE START OF THE FALL SEMESTER, THE REDWOODS were opened up for climbing. Meanwhile, Amanda LeBrun had sold their house in Corvallis and moved to California, where she and Steve rented a house.

"People have asked me why, in the end, I agreed to sell the house in Corvallis and go down to Arcata to be with Steve," Amanda LeBrun said to me. "There were so many issues in our marriage already. I had to give it the full shot. When I moved there to be with him, I knew it was a make-or-break time for us." She got a job teaching at a middle school.

Sillett liked teaching well enough, but he didn't think he was any good at it, and it was taking up huge amounts of his time. LeBrun helped him develop a curriculum for his courses at the university. "I almost never saw Steve that fall," she said. "He'd be on campus all day. He'd come home for dinner, and then he'd go back to his office and work until two or three in the morning. He ran field trips to the redwoods with his students almost every weekend. He would invite me to come, and that would be our time together. It was the Steve Sillett Show." The winter rains arrived in October, and it rained steadily, week after week, and Arcata turned into a gray, cold, sunless place. It was far from cities, culture, and art, and, for LeBrun, from nourishing human relationships. As for having children with her husband, that seemed to be out of the question. Amanda LeBrun became severely depressed.

"Writing was a lifeline for me," she said. Finding words for her unhappiness brought her something that her husband could not provide. She sensed that he was unhappy, too. She asked him if he would be willing to go to a therapist to get some counseling. He said that he didn't need any therapy. He said he was doing fine.

STEVE SILLETT'S CLOSEST FRIEND WAS HIS BROTHER, SCOTT, WHO WAS at Dartmouth College at the time, working to finish a Ph.D. in ecology; just as he had promised to Poe, he was embarking on a career to study migratory songbirds. He could talk with Steve both as a brother and as a fellow scientist. He flew out to visit Steve, and they joked and laughed, and spoke their private language—the language from their childhood that sounded like bizarre German. Even so, Scott found that his younger brother was becoming abrupt, fierce, intolerant, humorless, impatient with him and with everybody, and unwilling to talk about anything personal or intimate. Steve seemed to be in the grip of an agony that he couldn't express in words.

Steve had become afraid of flying to the East Coast, and he

stopped visiting home. "It's irrational, but I've had this fear that whenever I went back East there was a chance I would never return," he said to me. "I was afraid that I would die some pointless death on a city street." Then he stopped calling home completely. He even stopped talking with Scott, and the two brothers lost contact with each other. Steve's family felt that he had dropped off the face of the earth. In a sense, he had.

ONE DAY THAT FALL, EARLY IN THE CLIMBING SEASON, STEVE SILLETT, along with his friend and fellow climber Scott Altenhoff, hiked into the Atlas Grove carrying a backpack full of rope and climbing gear. While Altenhoff waited on the ground, Sillett made a solo first ascent into the Atlas Tree, which Michael Taylor had discovered five years earlier. No one had ever entered Atlas.

"At a hundred and ninety feet, Atlas splits into four huge trunks," Sillett told me. "In the center of the trunks is a crotch that contains a layer of canopy soil that we later measured to be one meter deep." It was a garden in the sky containing tons of dirt, along with sheets and beds of ferns, and thickets of huckleberry bushes. The canopy soil has been accumulating in Atlas for unknown numbers of centuries. It is composed of a mixture of rotting redwood needles, twigs, the roots of plants, and dust from the sky. The soil is apparently being fertilized by rotting lichens and twigs and redwood needles. No canopy scientist had ever before seen such a large amount of earth sitting so high up in a tree. "Michael Taylor named the tree well," Sillett says. "When you're up in Atlas, you get this overwhelming sense of a tree holding up the earth."

Sillett began to look at the other major trees in the grove, and during the coming months he would climb them one by one. He usually climbed solo, exploring the Atlas Grove by himself.

The one person who seemed able to maintain contact with Steve Sillett was Michael Taylor. Taylor was a loner who didn't expect much from people. He couldn't stand it when people judged him, and he wasn't about to judge anyone else. He asked Sillett if he needed any help with his climbing—help from the ground, of course. Taylor began coming along to watch and to carry gear. He would lean back against a log and stare at Sillett through binoculars as he went higher and higher.

Eventually, Sillett marked out a narrow plot of ground near the center of the Atlas Grove that he planned to study intensively. The plot covered exactly one hectare. A hectare is ten thousand square meters (a square that's one hundred meters on a side), and it's equal to two and a half acres. This was a long, thin hectare, laid out in a strip. He planned to study every tree inside the two-and-a-half-acre strip, large and small, regardless of the species. He would make a complete, three-dimensional map of the Atlas Grove plot, and he would attempt to identify all the living things in the trees. This was to be the first descriptive natural history of the redwood canopy.

As Sillett explored the giants in the Atlas Grove and got a sense of their characters as individuals, he gave them names. He named them mostly after Greek gods, especially the Titans. Starting at the western end of the grove, the principal redwoods in it are Zeus, Rhea, Kronos, Demeter, Epimetheus, Prometheus, the Seven Pleiades, and Atlas. Beyond Atlas grows a titan that Sillett named Ilúvatar, after the creator of the universe in Tolkien's *The Silmarillion*. Ilúvatar is the largest tree in the Atlas Grove, and one of the largest trees on earth. Beyond Ilúvatar stand Broken Top, Ballantine, and Bell. Michael Taylor had originally named this last one Bell-Bottom because the tree's flaring base reminded him of bell-bottom pants, but Sillett asked him if the name could be a little more dignified.

Tree climbers call a tree that has never been climbed a wild tree. A solo first ascent of a wild redwood is a hazardous operation. The crown of a redwood can bristle with rotting extra trunks. These dead trunks can be far larger than any tree in the eastern United States. They can be unstable, teetering, balanced on a mush of rotting wood, and they can collapse. The crown of an old redwood can be crisscrossed with dead limbs that may be several feet in diameter and hanging by shreds. Suspended dead branches are called hangers, or widow-makers. The twitching movement of a climbing rope can stir loose a widow-maker, and a falling branch can tear off other

CITIES ABOVE. A profusion of extra trunks and limbs is revealed in the naked, dead top of this smallish redwood. This gives a hint of the physical complexity of the living, foliage-shrouded, mazelike canopy Sillett hoped to explore in the Atlas Grove. *Drawing by Andrew Joslin.*

branches, triggering a cascade of spinning redwood spars the size of railroad ties coming down through the tree. A piece of a redwood can fall twenty to thirty stories down on a climber. A falling branch tends to get snagged on a climbing rope as it falls. Once it's caught on the rope, it slides down along it, heading for the climber. A falling redwood branch can spike itself five feet vertically into the ground. Sillett knew that a large falling branch, sliding down his rope for twenty or thirty stories, would tear through his body cavity and disembowel him. His thinking went like this: you are a grape hanging on a vine, and a falling branch can pop you.

The trees in the Atlas Grove had pieces of trapped dead wood in them that were bigger than Chevrolet Suburbans. As he climbed in the Atlas Grove, Sillett began carrying a small folding saw with him, and he used it to cut away any small, hazardous dead branches, but for the big dead structures he had to rely on hope. He found places in the Atlas Grove where entire sections of trunks had split away and collapsed. He referred to it as redwood calving. When a redwood calves, it gives off a roar that can be heard for miles, and it leaves a detonation zone at the base of the tree called a debris ring—a ring- or crescent-shaped pile of debris—and it looks as if a tank battle has been fought there. A calving event would obliterate anyone in its path.

Michael Taylor worked as Sillett's ground man. He helped Sillett carry ropes and manage gear. Taylor had sharp eyes, and he could spot problems in the redwoods. He watched for things wobbling or shaking: craggy-looking, unstable branches, hanging widow-makers, which might be concealed near the top of the tree, and shadowy dead structures hidden in the crown.

Sillett began to develop new techniques and rope systems for climbing in giant trees. Perhaps the most important of these is a complicated rig of ropes and carabiners, sixty feet long, known as the motion lanyard. It is also called the double-ended, split-tail lanyard. I think of it as a spider rope.

The spider rope works along the same principle as Spider-Man's silk. A skilled tree climber can anchor the ends of a spider rope over branches, and then move horizontally, vertically, or at a diagonal, through the air. You can hang from one end of a spider rope, like a spider hanging on a silk thread. You can also hang from both ends, suspended at the lower point of a V of ropes. You move through the tree by attaching alternate ends of the spider rope to branches, going from branch to branch, traveling on a succession of anchor points. Sometimes you swing on a spider rope to get somewhere.

Expert tree climbers also use another rope, a short one, in combination with a spider rope. This short rope, which is critically important for tree climbing, is called a positioning lanyard. A climber uses it much in the way a monkey uses its tail. The climber can get the lanyard looped around a branch and can hang from it, or can use it to pull himself somewhere. By using a spider rope in combination with a monkey-tail-like lanyard, a climber can move anywhere in midair in a tree, as long as there's something to hang from. This is known as full 3-D movement. A good tree climber can make 3-D movement look almost like weightless floating. The spider rope and positioning lanyard are the ultimate tools for exploring the redwood forest canopy. Sillett taught himself how to use them.

THE ATLAS GROVE WAS A NEW WORLD. SILLETT WANTED TO KNOW THE biomass of the grove—its total weight as a living system. He wanted to know how the mass was distributed in space—where the trunks and branches were. He wanted to know the shapes of the individual trees. He wanted to know the overall structure of the canopy in the Atlas Grove. He wanted to identify what lived on the trees, and where and how those things lived. How much rain do the trees get? How much sunlight? What's the surface area of their foliage? Where do the sun and the water go, and how are they being used? He wanted to understand the flow of water and energy through the Atlas Grove. The living and growing part of a tree is a very thin layer of wood, called the cambium, which exists under the bark, outside the heartwood of the tree. He had no idea of the actual size of the cambium of any giant redwood, and so therefore he didn't even know the actual size of the living part of any redwood organism. The whole grove was alive; it was consuming energy and growing; it was a fantastically complicated thing.

A person hiking through Prairie Creek Redwoods State Park could probably pass through the core of the Atlas Grove in five minutes of hiking at normal speed. Yet the exploration of the Atlas Grove—the Atlas Project—would be an arduous undertaking, one of the more ambitious projects in the history of forest-canopy science.

The remaining primeval redwood forest on the North Coast contained what Sillett calculated was twenty billion cubic meters of unexplored tree space. Twenty billion cubic meters of wild nature almost entirely unseen by human eyes. The exploration of a minuscule portion of it, just the Atlas Project, would push climbers to their limits. People make mistakes, and in tree climbing, as Sillett now understood, a small mistake can lead to death. The Atlas Project would require years of work, years of climbing, and more than one person who possessed expert climbing skills. Trained climbers with a strong knowledge of botany, zoology, and ecology would be needed.

Sillett had gotten to know a man named Robert Van Pelt, an ecologist who studies giant trees in Pacific Northwest rain forests. Robert Van Pelt is a large man with red hair and a red beard and a fierce obsession with trees. As a college student, Van Pelt wanted to be a physicist, but one day when he was sitting in a physics course he found himself staring out the window of the classroom into the crown of an oak tree, and he thought, I don't want to do physics anymore. With a bachelor's degree in physics, he got a job as a cook in Sequoia National Park. ("I could make hash browns in any shape from Mickey Mouse to Popeye," he told me.) The point of the job, as he saw it, was to keep him near the giant sequoia trees. Eventually, he got a Ph.D. in botany and he began studying trees, and then he began bashing around in dense areas of virgin rain forest in British Columbia, Washington, and Oregon, looking for giant trees and measuring them. Robert Van Pelt is responsible for discovering some of the world's largest and tallest trees of species other than the redwood—world-record Douglas-firs, red cedars, and Sitka spruces, among others.

Sillett taught Van Pelt the redwood climbing technique, and Van Pelt began climbing with Sillett, making measurements of the Atlas Grove and analyzing the data. Sillett also began taking on graduate students—master's degree candidates—and he began training them to climb. He was sleeping less and less.

∎ ∎ ∎

AMANDA LEBRUN FELT THAT IF SHE AND SILLETT BOUGHT A HOUSE of their own things between them might get better, but they didn't have much money and they couldn't find anything affordable. Finally, they signed a contract for a house on Fickle Hill, in Arcata. They closed on the house just before Christmas 1996, and moved into it during the holiday. A month later, LeBrun told Sillett that she wanted a divorce.

Sillett was taken completely by surprise. He wept and begged her to reconsider. He told her that she had blindsided him. He asked her why, when they had just moved into their house, she would choose this moment to ask for a divorce. Her answer was that life with him had become unbearable. He was no husband in any real sense, and she felt that he was incapable of loving her. And, she told him, she wasn't in love with him anymore.

He insisted that he was in love with her. He asked her if she would agree to go into counseling in order to save their marriage. She pointed out to him that she had asked *him* to go into counseling and he had refused. She told him that she didn't think their marriage could be saved. She felt that all of his emotions had been focused on the trees, and there was no feeling or energy left in him for anyone or anything else. She felt that their entire marriage was all about Steve Sillett and all about trees, and that was it. "Steve is a wonderful person in many ways," LeBrun told me. "I wouldn't have given nearly ten years of my life to him if he wasn't. But he was so wrapped up in himself that if something wasn't about him right then it disappeared from his frame of view."

They put the house up for sale, but discovered that they couldn't sell it for what they had just paid for it. They had overpaid. And since they had no money, they had taken out a mortgage that was most of the value of the house. The house sat on the market for many months, but they couldn't lower the price because they wouldn't be able to pay off the mortgage. During most of that time, they continued to live in the house. Sillett told himself that the marriage hadn't failed, even though they had stopped sleeping together. They finally did go into counseling, but it wasn't doing any good. "It was a killing experience for both of us," Sillett told me. "We were sleeping in separate bedrooms. We'd get up in the morning, and she'd be there in the kitchen, making coffee. It was torture for the soul." Speaking not a word to LeBrun, he would drink the coffee and go off to climb a redwood.

Sillett was embarrassed that his wife wanted to leave him, upset with himself for giving her cause to leave, and angry at her for doing it. She was the only woman he had ever loved, his first true lover. He didn't tell anyone that his marriage was coming apart. He kept it a secret from his colleagues at the university and from his closest friends. Not even his brother, Scott, knew about it.

Sillett felt ashamed of himself for giving up on the marriage. He thought about how his mother had saved his parents' marriage and probably his father's life by forcing his father to go into detox. He felt that he should have acted as his mother had done, trying whatever he could to keep the marriage alive. In his heart, he considered divorce to be sordid. He had never heard of any friends of his family in Harrisburg actually getting divorced. Nobody got divorced. Married couples might fight bitterly, but they fought to the bitter end. His grandfather Charles Sillett, who was married to Poe, didn't always get along well with her. Pop-Pop Charles had been full of sayings, which he often repeated to Steve and Scott, and one of them went like this: "Wish in one hand and shit in the other. Now clap your hands together, and what have you got?" It described how Steve's marriage had turned out.

Sillett shaved all the hair off his head, as if in penance. He also shaved off his eyebrows. Dark circles appeared under his eyes, like those of Edgar Allan Poe. He started wearing grungy denim coveralls with a zippered front, the kind that auto mechanics wear. He tore the sleeves off the coveralls. With his bald eyebrows and shiny head, and his huge bare arms coming out of ripped coveralls, he looked like a bum or a convict. That semester, he flunked 46 percent of the students in his course on general botany, giving them all F's. This sparked complaints from students to deans at the university, but Sillett simply explained that the students he had flunked had been doing lousy work, that was all. His students wondered if he was cracking up.

. . .

MICHAEL TAYLOR BOUGHT A HANDHELD LASER RANGE FINDER TO speed up the process of searching for tall trees. It looked like a pair of binoculars, but it fired a laser. It was used by surveyors to measure distances. You bounced the laser off an object, and got the precise distance to the object. Taylor could estimate the heights of a hundred redwoods in a single day with the laser, which enabled him to walk through the forest and quickly eliminate trees that weren't so tall; he could zero in on the tallest ones and get more precise estimates of their heights. But a laser isn't perfect. There is only one way to determine the exact height of a tall redwood, and that is to climb up into it and run a measuring tape down it.

One day in 1996, Michael Taylor took his laser to a state park in Mendocino County, where he discovered a redwood that, at 367.5 feet, was two feet taller than any other known tree at the time. He phoned Sillett and informed him that, guess what, he had actually found the world's tallest tree. Sillett climbed it and confirmed the discovery.

Taylor was convinced that he hadn't actually found the tallest tree, and he kept bashing around in the forest with his laser. One day he went with Sillett to make the first climb of Gaia, the ancient titan with a mushroom top and a cave in its roots. When Sillett got up to the top, he looked around and saw two beautiful spires sticking up, half a mile away. He took a compass bearing on them and descended, and he and Taylor followed the line along the ground, hunting for the bases of the spires. They found two extraordinarily thin, tall redwoods that they named Pipe Dream and John Muir, which stand next to each other. Sillett decided to climb Pipe Dream.

On the day of the planned ascent of Pipe Dream, Sillett was accompanied by Michael Taylor and his partner from the Telperion climb, Scott Altenhoff, who had started working with him as a second climber on some of the tougher climbs. Taylor and Altenhoff both noticed that Sillett seemed tense and subdued that day. They didn't know why he had shaved his head and his eyebrows, but they didn't give it much thought.

Sillett fired a dozen shots at the lowest strong branch on Pipe

Dream, but the arrows kept falling short. The branch was twenty-five stories above the ground. Finally, he gave up and fired the arrow over a weak-looking bundle of branches sticking out of the bark—an epicormic spray. He set the climbing rope over the epi spray and got ready to climb.

"Are you sure you know what you're doing?" Altenhoff asked him.

Sillett stared at him. "*Fuck it*. I'm going up." He clamped to the rope and leaped up it, climbing fast.

Taylor stood next to Altenhoff, watching with binoculars. Sillett was getting smaller and smaller as he went up Pipe Dream. When he got close to the epi spray over which his rope was anchored, he stopped and hung just below the branches on the rope.

Taylor didn't like this. He got on the radio. "Steve. What do those branches look like?"

Three of them were dead, Sillett reported. Ready to break. The fourth one, the smallest, was alive. His rope was wrapped over the whole bundle of branches, dead and alive. It was not a good anchor.

"I think you need to get down from there, Steve," Taylor said.

"Fuck it."

Somewhat to their suprise, he stood up on top of the dead branches, staring into Pipe Dream. He was standing at 220 feet.

Altenhoff asked Sillett to come down. Sillett gave him a negative.

Sillett had tied one end of his spider rope to the epi spray, and now he began throwing the other end of it up along Pipe Dream, trying to reach a stronger branch thirty feet above him. He couldn't. He kept getting his rope tangled on a second epicormic branch, a little thicker than a person's thumb, that stuck out of the tree above him, halfway toward the stronger branch. He began cursing.

Taylor got scared. He had never heard Sillett cursing in a redwood this way. Sillett liked to curse, but for some reason he never used profanity in redwoods.

Sillett flung his rope over the tiny branch, and then began to anchor himself to it.

"You don't want to tie in only to that one branch, Steve," Altenhoff said. "If it tears out, you'll have a long plunge to the ground and nothing to stop you." With a growing sense of horror, Altenhoff saw Sillett clip himself to the rope he'd just anchored over the thumb-size branch. "Steve. Come down," Altenhoff said sharply.

Sillett had gone into radio silence. He released his rope from the dead branches below him. They saw him swing out into space along-side the tree, hanging from nothing but the little epicormic branch, twenty-two stories above the ground.

Taylor wondered if his friend might be trying to commit suicide.

The only thing I have to lose is me, Sillett thought.

TEARS

O N THE GROUND, TAYLOR AND ALTENHOFF KNEW THAT SIL-
lett was in some place of personal agony that they didn't
understand. He began moving along Pipe Dream, inching
his way up the rope that was hanging from the thumb-sized branch.
He got close to it so that he could see it more clearly, and he saw how
his rope was looped over it, pressing down on it with the full weight
of his body. He knew that it could pull out of the tree with any sudden movement. He needed to get his rope up to the stronger branch,
which was now fifteen feet above him. Moving with the utmost care,
he began throwing the other end of his spider rope upward, trying to
hook the strong branch. He kept throwing and missing. He didn't
dare throw hard, for fear that the epicormic branch he was hanging
from would break and he would go down.

"I know you can make this throw, Steve," Taylor said on the
radio. "I know you can do it."

Sillett finally went for it, and threw hard. The rope looped over
the strong branch, and he clipped in and was safe. He got to the top
of Pipe Dream. Later, as he was rappelling out of the tree, he passed

by the little epi branch that he had hung his life on. His shoulder brushed it, and it popped out and fell, spinning around and hissing as it disappeared into the depths. It sounded like Kevin Hillery.

MICHAEL TAYLOR AND CONNI METCALF WERE MARRIED ON VALENtine's Day 1997, in the Chapel of the Bells, in Reno, Nevada. The bride was given away by her brother-in-law, and the couple honeymooned in Reno. They didn't want to have the wedding at Jim Taylor's estate in Santa Barbara, because it wasn't their style. Michael had finally graduated from Humboldt State with a degree in engineering. He quit his grocery clerk's job, and got a job as a salesperson in a Radio Shack store. He and Metcalf moved into an apartment near Arcata.

One day that spring, Taylor got a call from Sillett. Sillett wanted to know if he could come along on another one of Taylor's exploratory trips into the forest. There were a number of small valleys and pockets of rain forest in the parks along the North Coast of California that Taylor had identified as being populated with tall or large redwoods, but he hadn't explored them, and there were many valleys that he hadn't yet visited. It wasn't perfectly clear that anyone had ever looked through these valleys for groves of giant redwoods.

They began exploring Redwood National Park. The park encloses an irregular stretch of the California Coast Ranges northeast of Arcata, on the drainage of Redwood Creek (which is more of a river than a creek). By the time the national park was formed, in 1968, about two thirds of the land in it had already been logged. The park was centered on the Tall Trees Grove, the area that Howard Libbey, the president of the Arcata Redwood Company, had promised National Geographic officials that his company would never log. The park snaked along Redwood Creek. In the seventies, conservationists began referring to it as "the Worm," because of its shape and relatively small size for a national park. The conservationists began to push for an enlargement of Redwood National Park, so that it would include the entire watershed that surrounded the Worm.

The watershed of Redwood Creek was a rain forest owned by timber companies—Georgia-Pacific, the Arcata Redwood Company, the Simpson Timber Company, and others. It included dozens of tributaries of Redwood Creek—Devil's Creek, Bond Creek, and Emerald Creek, to name a few. The tributaries ran through V-shaped canyons, notch valleys, which were crowded with redwoods. The notch valleys rose steeply to ridges and mountains. While the debate simmered, the rate of logging increased dramatically in the adjacent, unprotected tracts surrounding the Worm, as timber companies rushed to take out as many redwoods as possible while they still could.

Bulldozer roads, called cat tracks, began converging on the park, pushing down along the ridge lines and into the valleys. The cutting often focused in on the groves of largest redwoods, a timbering practice known as high grading—the crews would fell the biggest and most valuable trees first, coring out the hearts of the groves, while leaving less valuable trees for a mop-up operation later. Enlarging clearcuts appeared all around the national park. Timber crews worked twenty-four hours a day, using floodlights at night, in an effort to get all the redwoods out before the government bought the land.

In 1977, a U.S. congressman from California named Phillip Burton introduced legislation to increase the size of Redwood National Park, and President Jimmy Carter signed it into law in 1978. The legislation allowed the companies to remove any redwood trunks that were lying on the ground. In the last days before the law took effect, some timber bosses reportedly instructed cutting crews to go straight into the redwood forest and drop every redwood they possibly could. The result, in certain spots on the North Coast, is radiating lines of clearcuts that seem to fan out into the redwoods from central staging zones, which today are visible from the air.

The federal government purchased all the land surrounding the Worm—totaling 48,000 acres. By then, about 80 percent of it had been logged or was empty of redwoods. Even so, the addition to the park contained 9,000 acres of old-growth redwoods, mostly tucked away in the deepest ravines and notch valleys, where the crews hadn't been able to push cat tracks. A mishmash of alder trees and tiny redwoods and little Douglas-firs, mixed with brambles, grew on the logged land, surrounding and walling off the remnants of the ancient redwood forest, burying the last shards in thickets of deep brush.

Michael Taylor and Steve Sillett decided to try to find and explore these lost fragments of Eden. They hoped to find undiscovered groves of giant redwoods, which might be tucked away in the jumble of the old logging operations. No trails led into the notch valleys of Redwood National Park, which were now surrounded by hellishly thick recovering clearcuts where nobody went. It seemed that nobody really knew if any giant redwoods might exist in the remote areas of Redwood National Park, or possibly in the dozens of tiny, inaccessible valleys that run through the redwood state parks.

As they picked their way down sixty-degree slopes and through cracklike gullies filled with brambles, Taylor and Sillett sometimes carried packs weighing sixty to eighty pounds. The packs were full of Steve's climbing gear and ropes, as well as laser-measurement equipment. They struggled over terrain that was almost too steep to climb. When they got into old-growth redwood tracts, the forest floor seemed to dissolve into a tangle of fallen branches and debris that had rained out of the trees over centuries. They stumbled across shafts and pits leading down into underground masses of buried redwood. Downed redwood trunks, ten or fifteen feet in diameter, lay in heaps. They traveled together in the rain forest, since traveling alone there is a bad idea. A person trying to climb over a crisscross of trunks could slip into a gap, fall thirty feet down into the center, break a leg, and never be heard from again.

One day, they got into a winding notch valley that I will call Fog Creek Canyon. (It has another name, but I am using a pseudonym for it, in order not to reveal the location of the giant redwoods that ultimately turned up there.) As they moved downstream along Fog Creek, they got into places the loggers hadn't reached, into stands of big trees, crowds of redwoods of large girth and great height. They could hardly see anything. They often had to backtrack, hunting for passages that would get them around piles of trunks or through debris. Bushwhacking through Fog Canyon was like exploring a chain of caves, and they had to do a lot of crawling. Taylor measured trees with his laser when he was able to see a redwood's top. Late in the day, they came out into a bowl in the forest, but the trees seemed to go on and on. They were thirsty and drank water from the creek, but Sillett was too tired to take off his backpack. They decided to turn around and find a way out. They began climbing up an extremely steep slope, hunting for an exit route upward. They climbed until they reached the edge of an old clearcut.

Suddenly, Sillett fell forward on the slope and sank down onto his knees. He put his hands over his eyes and face.

"Hey, Steve, are you okay?" Taylor asked. He thought that Sillett might be dehydrated and was perhaps getting sick to his stomach.

"I don't think I'm going to make it, Michael." Then he started weeping.

"Hey, man—what's wrong?"

"Amanda wants a divorce." He kept his hands over his face—wouldn't look at Michael.

Michael knelt beside his friend and put his hand on his shoulder. "I'm sorry, man. I didn't know."

"She's leaving me."

"Gosh, it must be awful."

Months had gone by, and Sillett and LeBrun were still living in the house together, but now, finally, that day he hiked with Taylor, Sillett saw that it was over. Taylor was the first person he could bear to say this to. He uncovered his face; it was shiny with tears. "I didn't see it coming, Michael," he said. "I'm still in love with her. She tore my heart out and stomped on it." He was sobbing uncontrollably. He held his face in his hands, and then he clutched at the soil.

He's really a very vulnerable person, Taylor thought. The shiny skin on his head, the forehead with no eyebrows, the coveralls—he looked like someone who didn't respect himself.

Taylor knelt beside his friend. He helped him sit up, and then he put his arm around his shoulder. "That is a hard knock, I know. Give it some time, though."

"I'm never going to get over it, Michael. I know I have to get over it, but I can't. I'm stuck."

"You'll get through it. I know you will."

"She doesn't want to have my children."

"Don't worry, Steve. Don't worry, you'll be a father someday. I know you will."

After a while, Sillett stopped crying. He got to his feet and began climbing up the slope. A little farther on, though, he went down on his knees again and covered his face.

Taylor held Sillett and let him cry. Sillett pulled himself together a bit.

"There are times, Michael, when I'm up in the top of one of these trees, and I'm looking down. It's a long way down. The thing is, I'm not afraid." The fact that he was no longer afraid of falling, he said, made him afraid that he might actually fall. When he thought about the detonation zone of his personal life, an urge to jump came over him, a desire to feel a release from gravity. He could hear the sound of the air rushing past Kevin Hillery, until it began to sound as if it were rushing past him. "I'm up there, I'm standing on a branch at three hundred feet, and it's just so easy to think about unclipping my rope and stepping off into space. Do you know what I mean, Michael?"

"I know," Taylor said. "I get these urges to jump when I'm in some high place. That's why I can't go into the trees with you. I'm really afraid that I would jump if I got up in a tree."

"It just would be so easy and so fast. Just unclip and step off . . . and take a long dive into a dirt nap."

"Please don't talk this way, Steve. Please, Steve. I know how you feel. My marriage with Conni isn't perfect. We've had our own problems. I know it isn't easy. I've had some personal problems, too." He was thinking about the pearlike shape of his body, his inability to make a living—all the zigzags and dead ends in his life. "Sometimes I don't think a whole lot of myself, Steve. My dad has rejected me. Man, he's cut me off. It's a crazy, mad world we're in. Think about the trees, okay? Don't think about Amanda. Don't think about falling, Steve, okay? Don't think about yourself, man. Don't think about anything but the trees."

Taylor stopped speaking for a moment, and they looked out from the slope and across the tops of wild redwoods that filled Fog Canyon. "It's almost like they're sentient beings," he said. "Maybe one of the trees we're looking at right now is the world's tallest. Maybe we'll find it some day. You have to keep going, Steve. Never give up. There is so much more to explore, so much more waiting to be discovered. Let's just keep moving, Steve, okay?" He helped his friend get back on his feet, and the two men continued climbing up the slope.

4

LOVE IN

ZEUS

THE LOST VALLEY

Taylor and Sillett pushed into dozens of pockets and creeks in Redwood National Park, hunting for unknown giant redwoods and their associated groves. They went into Fog Creek, Emerald Creek, Devil's Creek, Tom McDonald Creek, Bond Creek, Elam Creek. They hiked into many nameless tributaries of those creeks. They bushwhacked up into Lost Man Creek, but didn't find much. Feeling that they had searched Redwood National Park and found nothing very impressive, they moved their explorations northward, into parks closer to the Oregon border. They explored a few sections of Del Norte Coast Redwoods State Park, but didn't find any particularly tall or large redwoods. They penetrated Jedediah Smith Redwoods State Park, which lies along the Smith River, twenty miles south of the Oregon border.

One day in early May 1998, Sillett called Taylor and said to him, "Let's go out and try to find some champion trees." They decided to go to Jed Smith for a day hike, and be home by the afternoon. They were intrigued by a complex of small valleys that appears on the U.S. Geological Survey's topographic maps just to the south of Highway

199, inside the park. No trails led into the valleys, and the terrain was a clog of redwood jungle. The rain forest in Jed Smith state park is exceptionally dense, among the densest rain forests anywhere on earth, rivaling those of the jungle mountains of Peru and the fjords of southern Chile. The interior of the park is a warren of tiny, steep notch canyons and gullies. The understory of the forest consists of virtually impassable thickets of huckleberry bushes and salmonberry canes and ferns and small trees. The salmonberry canes are covered with prickles, which over hours of scraping can turn a person's exposed skin into an ooze of blood. Visibility in Jed Smith can be poor to near-zero.

About a month earlier, and about a mile and a half into one of the valleys, Taylor and Sillett had discovered a redwood titan that they named New Hope. They had given it this name because they felt that it gave them hope of finding more titans there. Now they planned to push beyond New Hope Tree, deeper into Jed Smith, exploring higher up into the complex of valleys, to try to get into valleys where they had never gone. It seemed unlikely that anyone had gone there in many years, and they would discover, once they got into the valleys, that the U.S. government maps of the area are inaccurate, and could not be used for guidance. For all practical purposes, the center of Jed Smith was a blank on the map of North America.

They decided to make their push into the area on May 11. It was a Monday, and Taylor was supposed to work at the Radio Shack store that day, but he called in sick. Sillett picked up Taylor at his apartment early in the morning, and they drove north. They didn't bring any food with them, or warm clothing, because they assumed that they would be gone for only half a day. They also didn't bring a global-positioning-system (GPS) locator device, since GPS devices typically don't work in redwood forests. They didn't bring cell phones, either, because cell phones also rarely work in redwood forests. They parked in a turnout along Route 199, and they went into the forest, pushing southward and upward along a creek toward New Hope Tree.

For the first quarter of a mile up the creek, they had to crawl through underbrush on their hands and knees, sometimes lying flat on their stomachs and belly-crawling. They wormed under tight masses of huckleberry bushes, or they turned their bodies sideways and rammed through them. Taylor was holding his laser range finder in one hand and trying to keep it from getting wet, and Sillett was carrying a laser in a knapsack.

After an hour and a half of clawing up the stream, they had gone about a mile. They regarded this as very rapid progress. They arrived at a fork in the creek, and both drainages headed up into notch valleys, one heading toward the west, the other toward the east. They decided to follow the west fork. Sillett was wearing a long-sleeved shirt, but Taylor was wearing only a T-shirt, and his arms began to bleed. The valley narrowed, blocked by redwood trunks. They climbed over and around them, and arrived at New Hope Tree. It had taken them two hours to cover a mile and a half of known terrain. They rested underneath New Hope, drank some water, and consulted the map.

The map showed a knoll or peak about a mile above New Hope Tree, in the middle of a warren of tiny valleys. They decided to make their objective the peak, where they might get a view, and then turn around and go home.

They began pushing into unknown terrain above New Hope Tree. Two hours later, they had gone just three-quarters of a mile farther up the gorge, which opened into a basin full of Douglas-firs mixed with redwoods. It was getting on toward noon. Both of them were tired and hungry, Taylor particularly so. He weighed 220 pounds, and his knees were beginning to hurt. They stopped and debated whether to turn back. They consulted the map. The USGS topographic map showed the knoll, or peak, right where they were. But there was no peak. Instead of a peak, the land went *down* and formed a basin.

USGS topographic maps are constructed by means of aerial photographs. In some cases, the area is not also surveyed on the ground. It would be exceedingly difficult to survey the terrain in Jed Smith, because visibility in the forest is exceptionally poor. The aerial photographs can reveal the shape of the top surface of the canopy, but they can't reveal the underlying terrain. In a redwood forest, the landforms can differ dramatically from the shape of the top of the canopy.

Taylor, in his explorations, had often seen small marks on redwoods—cuts in the bark, splashes of faded paint—that had been left by timber cruisers, men looking for trees to cut. Timber cruisers' marks persist for more than a hundred years. The timber cruisers were the vanguard of explorers in the redwood forest (after the Indians), and they went through the forests mostly during the twentieth century. This basin, Taylor noticed, had no timber-cruiser marks on the trees. "I honestly don't think any people had been in that place in a very long time," he said to me. "I sometimes wonder if people had been in there at any time after the discovery of the New World by Europeans."

Taylor and Sillett put their map away, since it wasn't doing them any good, and crossed the basin in search of higher ground. Some two hours later, they reached the edge of the basin, where they discovered a giant redwood that they named Neptune. They wanted to turn around and go back the way they had come, but there were no landmarks, the map was useless, and they didn't know exactly where they were. Instead, they went upslope, deeper into the center of the park, beyond Neptune. "We need to find an exit creek that will take us out of here," Sillett said to Taylor.

They began crossing a rugged, up-and-down plateau, clad with rain forest. It dropped down into another small basin that didn't appear on the map, and they came to a saddle between two ridges. It was now afternoon, and they had been going for seven hours. They began to zigzag back and forth, trying to find some feature of the land that was on the map.

Michael Taylor was beginning to get scared. Steve Sillett was in better physical condition than he was. The air temperature was in the fifties, and he was wearing only a cotton T-shirt. If it got dark, and began to rain, the temperature could drop into the forties and they would be soaked. In that kind of weather, a person wearing wet cotton clothing can get hypothermia, which is a very serious matter. They were seven hours of hard bushwhacking from any sort of help— but they didn't know exactly in which direction to go. They hadn't eaten in many hours, yet they had been burning up large amounts of energy staying warm and moving through the underbrush.

Sillett became worried about Taylor. He noticed that Taylor had a crazy and potentially dangerous way of moving through redwood jungle. When he came to a big redwood log, he would climb up onto it and sit there, and then he would fall off the log, disappearing on the far side with a crash in the underbrush. You're going to break your leg, he told Taylor, and Taylor answered that you have to let your body go limp as you fall, and then you won't break anything. Sillett called the move the Taylor Flop. Even though Taylor was in worse shape than Sillett, he kept moving ahead of Sillett, doing flops and crashes over logs. He seemed unstoppable.

They arrived at a waterless gully, a slot chasm choked with underbrush and blanketed in forest. They pushed down into the slot and came to a huge Douglas-fir. When they measured the tree with their lasers and a tape line, they discovered that it was the largest known Douglas-fir in California. They named it Ol' Jed. The discovery lifted their spirits momentarily.

The chasm continued downward beyond the tree, and they kept following it, hoping that it would come out somewhere, but it was a trap. They were mostly crawling through bushes, or sliding over boulders, or climbing over piles of huge redwood logs. The gully didn't seem to be going anywhere useful. "I suggest we call it Ruthlor Gulch," Sillett said to Taylor. "This describes its ruthless and unforgiving nature."

An hour into Ruthlor Gulch, Taylor, whose knees were beginning to swell up, sat down and refused to go any farther. "We should go back," he said. "None of this matches what's on the map. I don't want to have to sleep underneath a log." They hadn't brought a flashlight.

"We'll never find our way back," Sillett said.

"I'm turning back anyway."

"Dude! We can't turn back. We're committed to a heinous bushwhack."

At this point, Taylor blew up. He called Sillett a fucking tree fanatic.

Sillett told him to speak for himself. They began yelling at each other. Eventually, they calmed down and agreed that they had better behave like gentlemen or they were truly going to be hosed. "Look, we have to come out somewhere," Sillett said.

They didn't. Ruthlor Gulch just went on and on, for three hours. During that time they covered about a mile. It was a mile of worming, crawling, bleeding, sliding, Taylor-flopping, and cursing. Taylor began to have more difficulty walking as his swollen knees became more painful to move. Finally, the gulch came out into a nameless creek, jammed with boulders and logs, wandering in an uncertain direction. The creek had water flowing in it. They began to crawl in the water, since the creek was too choked with brush for anyone to stand up in.

An hour later, they were still crawling in the creek. They began referring to it as Cocksmoker Creek. The sun began to set, and the air became chilled and they were soaked. Taylor got the shakes from exposure to the cold water. It was apparent that a cold night was coming on. They weren't carrying any matches. Sillett and Taylor had decided, long ago, that they would never light a fire in a redwood forest, under any circumstances.

Taylor, who was leading the way, came to a fallen redwood trunk that bridged the creek. He climbed above it and stood up, and saw that the creek had come out onto level ground. Directly in front of him was a curving wall of wood that blocked his view. The wall was thirty feet across. It was the largest redwood trunk he had seen in all his years of exploring the North Coast. "*Aieeeee!*" he screamed.

SILLETT WONDERED IF TAYLOR HAD FINALLY BROKEN HIS LEG, BUT then he saw the titan. They circled around it. It turned out to be, in fact, two monumental redwoods joined at the base—a twin tree— with a combined diameter of thirty feet. It has since been named the Screaming Titans by Steve Sillett.

They thought they had made a fine discovery, but the discovery had just begun. When they walked past the twin titans they emerged into a grassy glade. Patches of open sky were visible, and pools of water shimmered. Around the edges of the glade stood a ring-shaped colonnade of undiscovered redwood titans—trees of enormous size in terms of mass. They are, collectively, the largest redwood trees on earth. The grove exists at the bottom of a hidden, notchlike valley deep in Jedediah Smith Redwoods State Park. It was previously undiscovered.

The hair was standing up on the back of Taylor's neck. He and Sillett didn't know what to say to each other. They felt as if they had walked into a dream. The stars were beginning to come out, and Venus was up. The trees were outlined against a deep-blue dusk. Near the Screaming Titans they encountered two monstrous redwoods, which Sillett would later name Eärendil and Elwing. They waded through the pools of water, and approached the row of titans growing on the far side of the colonnade. They ran Taylor's measuring tape around the nearest one. It proved to be one of the largest redwoods ever to have been found; they would name it El Viejo del Norte (The Old Man of the North). Next to it grows a redwood that Taylor and Sillett named the Lost Monarch. In 2003, Sillett completed a scientific mapping project of the Lost Monarch, and found that it was the largest living redwood in the world. The Lost Monarch contains at least forty thousand cubic feet of wood. Its trunk is thirty feet across—it's wider than the General Sherman, the giant sequoia in the Sierra Nevada that is the world's largest tree in terms of volume and mass. The Lost Monarch does not contain quite as much volume of wood as the General Sherman, however, because the Lost Monarch's main trunk has a more tapered shape, while the General Sherman is more of a fat cylinder. A redwood titan that was later named Stalagmight grows near the Lost Monarch, and there is Aragorn, Sacajawea, and Aldebaran. There are others.

The Grove of Titans was previously unknown to park officials and biologists. The trees in the grove had undoubtedly been looked at once in a while over the years by the occasional bushwhacker, or earlier by timber cruisers looking for trees to cut, but nobody had understood how big the trees are, or even the fact that it's a grove. A number of trails were surveyed and constructed inside Jedediah Smith park during the 1930s by the park service, but none of the trails entered the Grove of Titans.

As he walked through the Grove of Titans for the first time, Taylor began crying.

When biologists visit the Grove of Titans today, they vary their approach paths, so that their footsteps won't create a visible trail on the forest floor. The exact location of the grove is known only to a handful of biologists, who climb the trees and study the ecology of the grove. They guard the knowledge of its location with the jealousy of a prospector who has found a mother lode.

Traditionally, the "largest" redwood in Jedediah Smith state park is the Stout Tree, which is a tourist attraction. It grows in the center of the Stout Grove, near the Smith River, close to a road and a parking lot. On weekends in summer, dozens of people can be seen walking around the Stout Tree and looking at it and taking pictures of it. "The Stout Tree isn't even among the top fifty largest redwoods at Jed Smith," Michael Taylor said to me one day. The date of Taylor and Sillett's discovery of the Grove of Titans—May 11, 1998—is known to some botanists as the Day of Discovery.

Taylor and Sillett got out of the park that day by bumming a ride from a photographer whom they found photographing the Stout Tree. He kindly drove them back to their car. At nine o'clock at night on the Day of Discovery, they were stuffing themselves on cheeseburgers at a Carl's Jr. in Crescent City. It occurred to Taylor, as he wolfed down his second double cheeseburger, that he was eating too much. When he finished his dinner, he made a vow to honor the discovery of the Grove of Titans by going on a diet. Taylor soon lost fifty pounds, and he became a trim, fit man, with well-developed muscles and no visible fat. In addition to being probably the leading discoverer of giant trees in the history of botany, Michael Taylor is also the discoverer of the Taylor Diet. "It's simple," he explained to me. "I realized I was eating a lot. So I stopped eating a lot."

Robert Van Pelt, the scientist who, with Steve Sillett, had been climbing and studying the trees in the Atlas Grove, is in his own right one of the leading discoverers of giant trees. He is the author of *Forest Giants of the Pacific Coast,* a book that describes some of the largest known trees of various species, including redwoods. One day I was driving along the California Coast Highway with Bob Van Pelt—we were going to look at the Atlas Grove together—and he said, in an offhand way, "In the history of botany in the twentieth century, there was never a day like the Day of Discovery, and there will never be a day like it again."

"Why is that?" I asked.

"Because there is nothing on earth like those trees left to be found," Van Pelt said.

He was wrong.

WINDIGO

IN JUNE 1998, A FEW WEEKS AFTER THE DISCOVERY OF THE GROVE
of Titans on the West Coast of North America, Marie Antoine
graduated from Oregon State University with a degree in envi-
ronmental science and botany. She and her college boyfriend, Ted
Eldon, stayed together. They lived in Eugene, where she got a job selling flowers in an outdoor market downtown. It was a pleasant life,
with a musical drift, and they spent many evenings going to rock con-
certs. Marie Antoine enjoyed cutting the flowers and arranging them
in metal cans at the market and talking to passersby, and a fair number of people around Eugene got to know her. She was a calm person
with brown hair that had slight highlights from the sun, green eyes
with a touch of blue in them, and a delicate, heart-shaped face—
a woman of striking beauty. Marie Antoine was always friendly, but
she had a reserved air, and you could tell that she kept some things to
herself. I do not think that it would have occurred to very many people that the flower girl was developing plans to become a forest-
canopy scientist.

Her father, Ronald, had become ill while she was in college. His doctors didn't know what was wrong with him. He decided to move to Santa Barbara, where the climate would be easier on him, and he rented an apartment there. Marie visited him in Santa Barbara during the winter and vacationed with him in summer at the Cottage on Treaty Island, in Lake of the Woods. He was no longer the handsome, dark-haired man who had spent hours digging in his flower garden. He had become gray and pale, and was having difficulty walking.

Ronald Antoine wondered to himself how much time he had left. He wanted to see his daughter become a scientist before he died, but he couldn't quite come out and tell her. They would speak on the phone twice a week, and he began dropping hints during their regular chats, saying things like, "I hope you're filling out those graduate school applications."

It annoyed her no end. "Let me do things at my own speed," she said to him. She began filling out applications anyway.

In the fall of 1999, she entered graduate school in botany at Oregon State, with a plan to study the lichens of the forest canopy. For her master's thesis, with a distinguished botanist named William E. Winner, she decided to do research into *Lobaria oregana*—lettuce lungwort.

Marie Antoine began doing research on Lobaria at the Wind River Canopy Crane, a construction crane situated in a grove of tall conifers in the Gifford Pinchot National Forest, in southwest Washington. The Wind River Canopy Crane is an expensive instrument, a sort of giant telescope for looking into the canopy.

In 1990, Alan P. Smith, an ecologist at the Smithsonian Tropical Research Institute in Panama, got an idea that a construction crane could be used to lift people into the forest canopy. He rented a crane and put it up in the rain forest, and eventually two canopy cranes went into use at the institute. Not long afterward, Jerry Forest Franklin, a biologist at the University of Washington, managed to persuade the U.S. government to put up the money for the Wind River Canopy Crane. It has a boom with a gondola on the end of it, and it can deliver a person to the outer tips of branches of trees, where he or she can reach out and collect samples and do experiments. A scientist doesn't need to be an athlete to ride on a crane.

Marie Antoine had given up rock climbing after her forty-foot whipper. But she loved riding in the gondola; it was like flying through the canopy. She would visit the crane for a week at a time, and stay in a little cabin there. She gathered specimens of lettuce lungwort from branches and brought the lichens into the cabin and spread them out on blankets on the floor. She ran experiments on the lichens to see how they took up nitrogen under different conditions—changing humidity, changing temperature. She felt that Bill Denison's numbers for how much nitrogen fertilizer the lichen was putting into the forest were very general, and she wanted to see what the lichen was doing in particular places in the forest.

Antoine thought of herself as a sort of geek. At twenty-four, she had no way of knowing how long her career in science might last, or how far she would go, or whether she would make any lasting contribution to human knowledge. The only thing she knew for certain was that even if she spent the rest of her life studying lichen, whatever she found out about it would be almost nothing in comparison to what remained unknown about the tall temperate rain forests of the Pacific Northwest. It was like exploring the Labyrinth in the darkness with one fingertip, feeling around. The other thing about it was that most of the Labyrinth was in ruins or had vanished. Ninety-seven percent of the old-growth forests of Oregon are gone.

In June 2000, there was a scientific meeting at the Wind River Canopy Crane, and Marie Antoine was invited to give a talk. She was nervous, because she had never given a talk to a professional audience before, and she thought that her findings might be controversial. She had come to the conclusion that there was very little Lobaria in the trees around the Wind River crane. The crane could provide access only to a tiny slice of tree space. She suspected that there were other places in the forest where Lobaria was extremely important. She needed to get data from larger volumes of tree space, but she wasn't sure how to do this.

After the talks, there was a barbecue, and she started chatting with two graduate students, James C. Spickler and Billy Ellyson, who were working with the forest-canopy researcher Steve Sillett. The two men were expert tree climbers, and they began talking about climbing tall trees with ropes. Tree climbing was not something she had ever thought of for her own research. Ellyson turned around and said, "Hey, Steve."

Sillett came over, and the group continued the discussion about tree climbing. Like a feather across her face, it brought back a memory of a balsam fir and a little girl with a comb stuck in her hair, and the smell of a lake in summers lost. She knew who Sillett was because she had read a paper of his on lichens. They talked briefly about Lobaria, and the men drifted away. The next day, she was surprised to see an e-mail from him in her inbox. "Anybody who's going to study Lobaria HAS to see it on rope in the canopy, not on a crane," he wrote—shouting in caps at her, rather awkwardly. He had recently discovered a small, forgotten patch of ancient rain forest in northwest Washington, and he was going up there to explore it. If she was interested, she was welcome to come along, and he'd show her how to climb a tree. Sillett was by this time a divorced man. Amanda LeBrun had moved back to Corvallis, taking with her the trailer that he'd lived in when he had first arrived in town. She ended up living alone in it and writing, and she took a lover, whom she eventually married, and began work on a novel.

MARIE ANTOINE DIDN'T THINK MUCH ABOUT STEVE SILLETT. SHE wanted to learn how to climb a tree so that she could move forward with her study of Lobaria lichen. She got into her old Volvo and drove to the Reed College campus. Sillett had driven up to Portland from Arcata in his pickup truck. They loaded some camping gear into the truck and drove to Lake Quinault, just outside Olympic National Park, to a stand of thousand-year-old Douglas-firs. The biggest ones had been alive there since the height of Mayan civilization, and they were wild, unclimbed. When they arrived, they discovered that the road into the grove had been washed out. They would have to carry the ropes and climbing gear for quite some distance. Sillett brought his bow and arrows. Antoine carried six hundred feet of black tactical assault line in a bag that she balanced on her neck, on top of her backpack. They went into the forest, through clouds of mosquitoes, and arrived at the base of a giant Douglas-fir. The tree was three hundred feet tall and as big as a redwood. Sillett had discovered it a month earlier.

He began shooting arrows into it, and he kept missing. After two hours of shooting, he finally got an arrow over a branch. He used the fishing line to pull a cord up over the branch, and then they hauled a rope over the branch by pulling on the cord. As they began hauling the cord, Marie Antoine discovered that she didn't have enough strength in her hands to grip it. She was embarrassed about this, and she wrapped the cord around her hands to get more leverage. Her skin was burned and torn off in places.

Once the rope was properly anchored, Sillett spent five minutes giving Antoine a lesson in tree climbing. "Just follow me up when I call you on the radio," he said, and he disappeared up the trunk, ascending the rope.

Two hours went by. Sillett had gone into radio silence.

Antoine waited, not knowing what was happening. She looked at her hands and saw that she had mangled them, and wondered if she'd be able to climb. She set up her tent near the tree and went inside to spread out her sleeping bag. Finally, her radio beeped. "You can get on rope," he said, or something like that.

She put on a climbing saddle and a helmet, clamped a pair of ascenders to the rope, and started moving upward. Though she had done a lot of rock climbing, she had never used ascenders to climb a rope before. She associated hanging on a rope with a near-death fall, and she was frightened in some deep way.

At 150 feet—halfway up the tree—Antoine had a panic attack: she couldn't climb any higher. In the past, when she climbed a cliff she had always had a rock face to grip with her hands and feet, and she could look at the rock rather than looking down. Here, there was nothing but space all around, except for this immense column of wood, and she was hanging in the air away from it, and couldn't get a grip on anything.

She thought about calling Sillett on the radio and telling him that she was scared and asking him to lower her to the ground on the rope. You can either bail and never climb a tree again, or you can just go up, she told herself. She was still trying to decide whether or not to call for help when she found herself inching up the rope. By the time she reached the top, thirty stories above the ground, she was physically exhausted. She sat down on a branch, got her arms partway around the trunk, and held on.

The view was incredible, looking across the lake and up to the snowcapped peaks of the Olympic Mountains. "I'm glad there are no mosquitoes up here," she remarked to Sillett. The canopy is stratified into layers, with different organisms living at different levels, like different species of fish living at different depths in the sea, and mosquitoes rarely live near the top layer.

"Go ahead and try standing on the branch," Sillett said. He showed her how to balance on a branch while keeping most of her weight suspended on a rope.

She wasn't sure that she could get up and just stand on a branch. She almost lost it, and nearly told him that she had to go down. Struggling with the rope, she got herself into a standing position, and then sat down on the branch again. Then they had a highly technical conversation about Lobaria, and that made her feel better. The sun started to go down, and they watched it in silence. He began stringing up a Treeboat. He planned to sleep in the tree. "This tree doesn't have a name," he said, as he worked. "Would you like to give it a name?"

She thought for a moment, and answered, "I think its name should be Windigo."

"What does that mean?"

She explained that the windigo is an Ojibwa spirit that is said to have killed many people around Lake of the Woods. The windigo emerges from the forest at night, near a person's tepee, and begins whispering. If you are overwhelmed by the call of the windigo, and go outside into the darkness to meet it, you will be destroyed, leaving nothing but a few drops of blood on the ground.

Sillett thought it was a gruesome name to be associated with a tree.

There was more to the legend of the windigo, she said. Her father and mother had told the story to her and her sister as if it were a fact, and she had been absolutely terrified of the windigo as a child. Realizing that they had gone too far with the story, her parents made up a happy ending. They told the girls that if you resist the call of the windigo and stay inside the house, and listen to the windigo's call without going outdoors, you assume the windigo's power. You take the force of the windigo inside you, and you can do great things.

NIGHT CLOSED IN, AND ANTOINE RAPPELLED OUT OF THE TREE BY herself. She curled up in her sleeping bag in her tent and read a book with a flashlight. She was too tired to sleep. She looked out of the tent and saw that there was no light shining at the top of the tree.

Sillett had a particular way of sleeping in a tree. He would clip his rope to a branch, so that he couldn't fall, and then he would lie on his back in the hammock and crisscross his arms at diagonals over his chest, like a pharaoh lying in a sarcophagus. At the same time, he would splay out his legs and place the soles of his feet together, flat against each other, like a frog. It was a sort of yogic position, maybe the Frog Pharaoh. He never had any sensation of falling asleep: morning would come an instant later. He never dreamed in trees.

When day broke, they talked on the radio about what they would do next. Sillett was taking measurements of the top of Windigo, and she told him that she was coming up to help with the work. She began ascending along the rope, but thirty feet above the ground her right hand cramped up and the fingers froze into a sort of claw; she couldn't unlock them. She had overworked herself climbing the day before, and the burns on her hand probably didn't help. The cramp wouldn't go away. She called Sillett on the radio. "My hands are sore." She told him that she was really sorry but she had to bail out. He came down from the tree and they packed up the ropes and headed home.

They traveled for hours in the pickup truck, and they talked more about lungwort. At some point, Antoine smoothed her hair or did something with her hands, and Sillett noticed the burns. "What did you do to yourself?" he said. "You've torn the skin off your hands. Oh my God, Marie, those are massive burns."

"It's nothing."

He was horrified that he hadn't noticed her hands. How could he have let her hurt herself like that and not seen it? Most people with burns like those would have insisted on going to an emergency room, but she hadn't even mentioned it. What kind of woman was she?

Antoine tried to calm him down. It hadn't been worth mentioning, she said. Her stupidity in getting her hands burned, and then getting one of them so cramped that it looked like a claw, made her feel incompetent. He dropped her off by her Volvo on the Reed campus, and she thanked him and turned away to get into her car. He offered to give her a hug, and it became an awkward moment, because she wasn't expecting it. She turned around and, with stiff arms, held out straight, gave him a light squeeze.

She got into the Volvo, and it rumbled to life and clattered off. The car was falling apart.

SILLETT THOUGHT THAT THE "HUG" WAS A MESSAGE. IT MEANT THAT any relationship between them was going to be strictly professional. She was Canadian, apparently. She liked to visit her father, who wasn't well, in Canada, and her mother was no longer alive—that was all he knew about her.

Fall arrived, winter came. Sillett told himself that he wasn't interested in Antoine in a personal way. He sent her an e-mail, asking her how her research into Lobaria was going. She replied, inquiring about his research. His research was going quite well. He wrote to her about the Atlas Project. He was making a huge, three-dimensional map of the trees in the Atlas Grove, and he had plans to start making a 3-D map of the Grove of Titans, too.

One day Sillett was climbing in the Pleiades—a cathedral complex of seven trees and trunks in the Atlas Grove—and he had an accident. He slipped and fell out of the top of Pleiades I, which is 310 feet tall. Dropping down through the air, he reached out and caught a branch with one hand. This ripped his shoulder out of its socket and tore a piece of flesh out of his hand, but it also stopped his fall. He ended up hanging from the branch by a bleeding hand and a dislocated shoulder, twenty-eight stories above the ground, and feeling a bit surprised with himself. He mentioned the incident to Antoine in an e-mail.

. . .

MARIE ANTOINE'S FATHER BECAME VERY ILL, AND HIS DOCTORS
finally diagnosed his condition as a rare disease called polyarteritis.
By the time they caught it, it was past the point where his life could be
saved. Marie and her father had both sensed that he might be dying,
and now there was a name for his disease and a timeline. It was one
of the deepest disappointments of her life. She had clung to her father,
and had been looking forward to having a relationship with him as he
grew older. This was not to be. She did not mention anything about it
in her e-mails to Sillett.

She was spending most of her time working in a laboratory on the
Oregon State campus, where she was putting bits of lungwort into a
plastic tube and measuring how much nitrogen the lungwort took
from the air as she varied the temperature and humidity inside the
tube.

It was in the laboratory, not at the Wind River crane, that Antoine
finally understood the scope of her research. She realized that there
was a lot of data on the climate of the old-growth forests of Oregon,
and that she could combine this climate data with the data she had
gathered from samples from the Wind River crane, and thus develop
a model for how much nitrogen the lungwort was creating in different habitats. It was going to be a lot of work, but it might provide
some powerful insights. "I realized I could build a really simple
model, because nature is simple," she said to me. "That was so cool."

William Denison, in his pioneering work on Lobaria, calculated
that every year, on average, Lobaria puts two kilograms (four and a
half pounds) of nitrogen fertilizer into every hectare (two and a half
acres) of old-growth Douglas-fir forest. This was not a huge amount
of fertilizer, but it was very important in the ecosystem, and it was a
self-sustaining process that originated in the canopy, in space above
the ground. It was a new look at life in a forest. Marie Antoine discovered that in places where Lobaria is abundant, and where the temperature is cool and the humidity is just right, it can produce up to
five times as much nitrogen as Denison had estimated. Antoine's work
revealed that in some places Lobaria is introducing up to seventeen

kilograms (almost forty pounds) of new, usable nitrogen into each hectare of the forest ecosystem. It all ends up as fertilizer. It's like showering a sack of fertilizer in a little patch of forest each year. In other spots, she found, Lobaria is hardly doing anything. She had developed a clearer picture of something that was happening in nature, and in finer detail. It was like looking at a planet with a small telescope and seeing something blurry, then looking at it with a better telescope and seeing mountains and plains.

"Lobaria's a type of lichen that spreads from place to place when pieces of it break off," Antoine explained to me one day. "Some of them are big, chunky pieces, but the lichen also has tiny, frilly edges that are designed to break off." Lobaria spreads through a forest canopy during storms. A big storm will tear bits of the lichen off branches, and the wind will carry them for considerable distances. The lichen spreads through the crowns of old trees like an infection.

It is a slow-moving infection. A piece of Lobaria the size of a child's hand might take ten years to grow to that size. (Lobaria is a comparatively fast grower. Some lichens can take twenty years to become the size of a dime.) It can take years or decades for some species of lichens to spread from one tree to the next. "If a whole mountainside has been cut, it will be a very long time before the Lobaria comes back," Antoine said. "You start to see it after about two hundred years. But you don't see big, juicy, drippy abundances of these lichens for centuries. You only see it now in old-growth Douglas-fir forests that are over five hundred years old."

A stand of Douglas-firs may be three hundred years old, older than the United States of America, but it will still be a young patch of forest, devoid of many species of lichens. A stand of trees in a temperate Pacific Northwest rain forest that began growing at the time of the Magna Carta (1215) will only now be reaching a fullness of biodiversity. It will be loaded with a variety of lichens and mosses that don't occur in younger forests, and it will also contain a much greater variety of animal life, large and small.

Marie Antoine's research reminds us that the old-growth Pacific Northwest forests that have been logged away in recent years cannot return to a climax, old-growth state until A.D. 2500 to A.D. 2800, even if they are left alone. In the next five to eight centuries, no one knows

JEWEL OF TIME. *Lobaria oregana*, or lettuce lungwort, growing on the branch of a thousand-year-old Douglas-fir. *Drawing by Andrew Joslin, after a photograph by Thomas B. Dunklin.*

what will happen to the earth's climate or the conditions of life for people, but the great Pacific Northwest forests that are gone will not be seen again anytime soon.

AROUND THE TIME THAT MARIE ANTOINE AND STEVE SILLETT WERE exchanging messages about lichens and maps, a man named Chris K. Atkins, who lives in Sonoma County, California, developed an interest in tall redwoods. He eventually heard about Michael Taylor, and they became friends almost immediately. Atkins bought himself a laser range finder, and he and Taylor began exploring the redwood forest together, operating frequently as a team.

Chris Atkins is married and has a child, and he supports his family by paying visits to fast-food restaurants all over California, where he leaves out packs of advertising coupons. The coupons tout hot deals for condominiums for sale in resort communities. People can fill out a coupon with their name and address on it, and put it in a box. Atkins collects the coupons and sells them to resort communities as sales leads. "The truth is, I'm obsessed with redwoods," he said to me. "My work has given me the freedom to find them and study them." His colleagues in the coupon business don't quite get it.

Atkins and Taylor began to survey the redwood forest intensively, hitting every tall-looking tree with their lasers, and they began discovering many more supertall redwoods. Working both as a team and separately, they discovered Flood Line, Mosque, Obsidian, Pig Snout, Pinnacle, Bamboozle, Wounded Knee, Tranquility, Cloud Nine, Obelisk, Tosca, Tenador, Trifecta, Dome Top, Lone Fern, and the Radford Stovepipe, to mention a few. By 2000, about a hundred redwoods of the tallest class, trees more than 350 feet tall, had been found, including one that Atkins named Millennium, because they found it a few days after the turn of the millennium.

Chris Atkins saw that he and Michael Taylor were accumulating some very interesting data. The world's tallest trees are organisms that live at the edge of what is possible in nature—they push the limits of what a tree can do. For this reason, they are very sensitive to any changes in the climate or the environment. The tallest redwoods may be reacting to global warming. They may also be reacting to the steady increase in carbon dioxide in the atmosphere that has occurred as a result of human activities—from the burning of fossil fuels, for example, and from the burning and destruction of rain forests. Trees take up carbon dioxide from the air, and they release oxygen into the air. When a forest is burned down, the carbon in the wood ends up in the air as carbon dioxide, and there are fewer trees to take up the carbon dioxide from the air afterward.

Chris Atkins was measuring the tallest redwoods regularly, going back to them and measuring them again, watching them change. He began to get a feeling—with data to back it up—that the redwood forest as a whole might be getting taller, like a rising tide. Why? Was this a growth spurt of a whole forest due to increased warmth, increased carbon dioxide in the air? Steve Sillett realized that the coupon man was doing important research, and he invited Chris Atkins to join him as a co-author of a scientific paper.

Michael Taylor also had a new career. He got a job as a mechanical engineer with the Alza Corporation, a pharmaceutical company in Davis, California. He had an office in the factory, where he was the Mr. Fixit guy for the machines that make Nicoderm antismoking patches. Whenever there was a glitch with the machines, day or night, the plant managers would call Taylor and he would fix it. He seemed to be the only person who could fix the Nicoderm machines when they had problems, and so people who were using the patches to quit smoking had developed a slight dependency on Michael Taylor, unbeknownst to themselves.

Taylor and Atkins, exploring the redwoods together, were finding so many new tall trees that they began to wonder if the world's tallest tree still hadn't been found. The tallest known tree was the Mendocino Tree, which Taylor had discovered in 1996. Taylor and Atkins began referring to the hypothetical world's tallest tree as the Crown Jewel. Sillett told them to forget it, there was no Crown Jewel out there.

One weekend in July 2000, Chris Atkins phoned Michael Taylor and asked him if he wanted to spend a day measuring redwoods in Humboldt Redwoods State Park. Taylor was preoccupied with some trouble with the Nicoderm machines, and Atkins went by himself. He carried a tripod with him so that he could keep his laser steady while he measured the trees. Around noon, he found himself in a stand of redwoods that was unfamiliar. He had a feeling that no one had ever looked at these trees carefully before; Michael Taylor had never mentioned this grove. He walked around for a minute or two, and then, holding his laser device in his hands, he looked through the eyepiece and aimed the laser at the top of a very big redwood, bathed in summer light. The readout was rough. It showed that the tree was around 366 feet tall, within a foot of the Mendocino Tree. Oh, my God . . . , he thought. He set up his tripod so that he could get a more accurate reading. The top was moving in a breeze. He waited for a minute until there seemed to be a pause in the tree's restlessness, then he aimed the laser beam at the topmost sprig of foliage. This time he got 368 feet— six inches taller than the Mendocino Tree. Chris Atkins had found what might be the Crown Jewel. He packed up his tripod and moved on, feeling a quiet sense of satisfaction. He named the tree the Stratosphere Giant.

Later, Sillett and a graduate student, James Spickler, made the first climb of the Stratosphere Giant. They measured it and confirmed that it was the tallest tree. The Stratosphere Giant was growing by two to five inches a year. In 2005, it passed 370 feet, with no sign of stopping. It is thought to be roughly two thousand years old.

Sillett e-mailed Marie Antoine about the discovery. Atkins, the coupon man, had discovered the world's tallest known living thing.

Marie Antoine was puzzled by Steve Sillett's relationships with other people. He seemed to collect really odd characters around him, she thought. It made her wonder about *their* relationship. The situation was rather Victorian. She was exchanging letters with a fellow botanist, and the letters centered on a mutual passion for lungwort. The letters were correct, polite, and slightly formal.

LATE IN 2001—A YEAR AFTER ANTOINE HAD LAST SEEN SILLETT—SHE drove down to California to spend a weekend climbing a redwood in the Grove of Titans with him. She borrowed a car from a friend (she didn't think her Volvo would make it to California) and said goodbye to her partner, Ted Eldon, for the weekend. It would be her first time climbing a redwood.

Sillett had begun climbing and collaborating with George W.

Koch, an ecologist and plant physiologist at Northern Arizona University in Flagstaff. Sillett had flown to Flagstaff to go over some research they were doing, and while he was there he had told Koch about Marie Antoine. "She's beautiful, George," he said. "And listen to this: *she's into Lobaria.*"

Koch had told him that sounded fairly unusual.

Now, in Arcata, as a climb with Antoine in the Grove of Titans loomed, Sillett began to go to pieces. He telephoned Koch in a panic. "I'm in a sad position," he said. "I'm feeling truly worthless. It's obvious Marie doesn't like me. She's just interested in climbing redwoods."

"Come on, Steve. Do you think all she's interested in is climbing trees?" Koch said.

Antoine and Sillett met in Arcata and went to dinner at a sushi restaurant, where they talked about lichen. Then they went back to his place. It was a clean, empty cave of an apartment that reflected the emptiness of his personal life. He had reduced his possessions to a small number of items, mostly clothing and books that would fit inside one suitcase. The kitchen had nothing in it. He was living on sushi, whole-grain muffins, and Snickers bars. There was no desk, no chairs anywhere in the place. Linoleum on the kitchen floor, bare carpet elsewhere. The rooms were lit with bare bulbs on the ceilings. In the living room, there was a futon couch and a television set and one low table. They sat on the futon and watched a video of people climbing redwoods. In it, Tom Ness, the co-founder of New Tribe, is crouched inside a thicket of huckleberry bushes at the top of a redwood pouring himself tea from a teapot. He starts flapping his arms and cawing like a raven.

Sillett offered Antoine the futon couch to sleep on. She spread out her sleeping bag on it and used her sweater for a pillow.

They got up the next morning before dawn and drove north to Jedediah Smith state park, where they hiked several miles into the Grove of Titans. They carried large amounts of rope and gear with them, and arrived at the Screaming Titans, which had never been climbed. Sillett wanted to get up into them and do some reconnaissance. Earlier, he had strung a line into them with his bow and arrow, and now he attached seventy-five pounds of camping gear to himself and ascended on a rope into the canopy.

When he got up into the twin tower, he didn't like the look of it. The Screaming Titans seemed healthy from the ground, but the tops were a rotted-out death trap, ready to collapse in a calving event. He radioed down. "We're going to need to transfer to some other trees after you get up here," he said to Antoine.

She wondered what that meant. She got on the rope and began moving up. Meanwhile, Sillett positioned himself facing Eärendil, the giant redwood that was closest to the Screaming Titans. He got a rope into Eärendil and skywalked over to it.

Antoine reached the top of the Screaming Titans.

Sillett rigged a horizontal rope called a zipline between Eärendil and the Screaming Titans, and he invited Antoine to travel over to him. She suspended herself from the zipline on a pulley and soared over to Eärendil. She was three hundred feet above the ground. She wasn't afraid, but she felt extremely wary.

The process of getting from one tree to another had been slow and quite difficult, and Antoine had lost all sense of time. The sun was going down. Sillett got busy setting up a campsite, tying two Treeboats to different places in the upper crown. Then, while there was still some light, they decided to go to the top of Eärendil for the view. They climbed thirty feet above the Treeboats. Marie Antoine decided to climb barefoot, and she took off her shoes and tied them to a branch.

He teased her. "That's kind of weird, climbing around barefoot. What are you, some kind of hippie?"

She said that she had gone around barefoot all summer as a child living on the island in Lake of the Woods, and the soles of her feet had been fairly tough ever since. He asked what the island was like, and she said she was going there soon to visit her father.

Eärendil turned out to have a cathedral-like top with eleven spires rising through it. They moved from spire to spire. Nobody had ever been in the top of Eärendil before. He showed her how to use a spider rope to branchwalk. She wasn't too bad at branchwalking barefoot. They branchwalked out among the spires and attached their ropes to branches, then they stepped out into space and descended about five feet, suspended in midair at the outer edges of the crown. They watched the sun go down over the Grove of Titans.

They ate a supper that consisted of avocados and freshly baked bread. The food was Antoine's contribution, and she showed Sillett how to properly eat an avocado when you're camping. The avocado has to be ripe and soft. You cut into it with a knife, making a small slit in the rind, and then you squeeze the avocado. The insides squeeze out through the slit onto the bread. It's like squeezing a tube of toothpaste. Neat, with no mess. They were extremely hungry.

As they ate the bread and avocado, they were looking down toward the pools of water that Sillett and Michael Taylor had waded through on the Day of Discovery. They were black in the evening light. Sillett pointed across the tops of the titans—El Viejo del Norte, the Lost Monarch, and Aldebaran. It was amazing to think that he and Taylor had found this grove just three years earlier. Close by, less than a hundred feet away, was Elwing, the wife of Eärendil, he said.

"Who were Elwing and Eärendil?" she asked.

"They were the parents of Elrond." Elrond, he explained, was the half-man, half-elf king in *The Lord of the Rings,* who protected Frodo and his companions in Rivendell. Elrond's daughter, Arwen, fell in love with Aragorn, a man. Though she had been born immortal, she chose to be mortal and to grow old and die with him. "Eärendil and Elwing and their descendants were given a choice: they could choose to be mortal and be released from the world by death, or they could be immortal and be tied to the world for eternity."

She told him about losing her mother to bone cancer when she was eight, and about her mother's attempt to stay alive long enough so that she could form memories of her. "The truth is that I can't remember much about my mom, and I think it must have been a defense mechanism," she said. "It was such a painful time." The small things that she could remember about her mother, though, were crystal-clear. Now her father was dying, she said, and this might be their last visit to the island together.

They got into their hammocks, and the stars began coming out. Antoine wondered about the human necessities before bedtime. Sillett swung around the side of the tree and unzipped his fly. Antoine wasn't certain how a woman is supposed to pee while she's wearing a tree-climbing harness. She sure didn't want to take the harness off and leave herself unattached to the tree. She wondered if she should ask about the procedure. It became imperative. "What do I do?"

"It's no problem, you just do it," he said drowsily.

It was a huge pain for a woman. He didn't realize how hard it was. She had to undo the main belt buckle, which nearly detached her entirely from the safety harness, and she struggled with everything, wobbling around on her rope, and the whole thing seemed anything but safe.

When she finally settled down into her sleeping bag in the hammock, the harness turned out to be uncomfortable. She found that she had to sleep on her back, with her legs elevated and her feet pointing up. She never slept this way at home; she always slept on her side. She could hear Sillett's breathing becoming slow and soft. He was sleeping like a baby. He was lying on his back, with his arms crisscrossed over his chest and the soles of his feet planted together and his knees splayed out like a frog. Strange.

She couldn't sleep. There was no moon, and it became almost pitch-black inside the crown of Eärendil. The stars were isolated pinpricks, visible through the foliage. Then she began to hear sounds. The redwood canopy came alive with sound. There were little chirps and squawkings, flutterings of wings, rustlings. What were they? Birds? Flying squirrels? Salamanders? Insects? She wondered if these living things were unaware that humans were present. Six billion people on the earth, and how many have seen the Grove of Titans? Fewer than ten people, she thought. How many have seen it from this tree? Exactly two people. I guess I'm a lucky human. . . . She had been asleep, because she suddenly woke up. The world was different. The stars were gone, everything was hushed, and she could smell the sea. A fog had flooded up into the valley and covered the trees. Eärendil became damp, and foggy dew began dripping on her face. It kept her awake. The sun rose through a layer of fog.

THE FOG BURNED OFF, AND THE DAY TURNED BRIGHT AND WARM. They spent the morning taking down the hammocks and removing their gear from the top of Eärendil. They rappelled out, one at a time, sliding down the rope and along the endless trunk down to the ground. When Sillett touched down, he felt a sense of release—a sense that nobody had been hurt and he hadn't made a fool of himself with her. Antoine touched down, and the normal world came back, and it felt complicated to her. They pulled the ropes from the tree and hiked out of the Grove of Titans. They reached the truck around noon. It was a Sunday, and Marie Antoine had to drive back to Oregon, but there was a little time left before she had to leave, and they decided to go for a swim in the Smith River, which runs through the park. As they swam, Sillett took off his shirt and talked about plants. His upper body was dripping as he came out of the river, the muscles defined and goosebumps all over his skin. Afterward, they walked along the beach near Crescent City. The surf was crashing and a cold wind was blowing, and even in the sunshine they could feel the autumn in the sea. The swim in the river had left Antoine feeling cold, and she started shivering. "I get cold really easily," she said. Without thinking, he started to take her shoulders in his hands, so that he could rub them and help her get warm. But he pulled back. He couldn't touch her—he hadn't touched her since that wretched moment a year ago when he had tried to hug her. He put his hands down. She rubbed her own shoulders, shivering, not looking at him.

She was afraid of what was happening to her. It was beginning to make her feel miserable.

Sillett had gotten a year off from teaching, and in a few months he was planning to go on a sabbatical trip to Australia, where there were huge trees that he wanted to explore—*Eucalyptus regnans*—the tallest trees in the Southern Hemisphere. They can be nearly as tall as redwoods. He wanted to begin exploring other supremely tall canopies. He wanted to go beyond the redwoods and explore the tallest forests on earth—in Australia, Asia.

She was silent as they walked on the beach and he talked about his dreams of exploring the earth. Afterward, they still had a little bit of time together, so they went to a coffee shop to get some lunch, where they ate fish-and-chips. He said to her, "If we do more climbing together, we could be a dynamic duo."

His words embarrassed her. They were so awkward. What is he trying to say? she thought. The phrase "dynamic duo" sounded weird, and what made her very nervous was that it also seemed right, in a way. What did he think of her, exactly? "Yeah, well, I guess," she said.

They said goodbye to each other on the sidewalk in front of the coffee shop. This time she took him by the shoulders and gave him an affectionate hug. She got into her car and drove north.

SILLETT WENT BACK TO HIS APARTMENT IN ARCATA IN A STATE OF TUR-moil. The hug was only the second time they had touched, and they had known each other for more than a year. He told himself that he was absolutely not falling in love with Marie Antoine. Not so. No way. He thought he might be falling in love with the *possibility* of falling in love with her. She had not reacted well when he said they could be a dynamic duo. And he had shot his mouth off about exploring forests in Australia. He had said too much. He could see different futures forking out from the present moment. In one of them they were lovers, and in the others they were sending each other the occasional e-mail about lichens.

As Antoine drove back to Oregon, she couldn't get the previous two days out of her mind. Each moment stayed in her memory, crystallized and permanent. She had maintained her honor. He had been perfectly honorable with her, too. They had not kissed, and he hadn't offered to kiss her. It was funny—she hadn't thought about kissing him, not even as a possibility. Nevertheless, their relationship had been intimately and powerfully physical. They had been so close to each other, suspended on ropes in the air, moving around each other in the deep canopy almost like two dancers, far out of sight of the ground. At certain moments they had depended on each other for their lives. It was as if they had made endless love in the air without touching.

It made her wonder about what love means, how to define it. She had loved Ted Eldon for a long time, ever since college, but what was the nature of her love for him? They were living together, they were lovers. She wished that Ted could be her brother, so she could love him like that, but he wasn't her brother. She wondered what she really knew about love. It occurred to her that, at twenty-four, she might be in love, really in love, for the first time in her life. It scared the daylights out of her. Or perhaps she was just torturing herself. She knew that she could love climbing redwoods, but she wasn't sure that she could love Steve Sillett. She tried to separate these two things in her mind, the trees and the man.

She wondered how it would be possible to be in love with him. He was a divorced guy. He was complicated and eccentric, and he had odd friends. He loved to talk about his friends: Michael Taylor, the billionaire's son; Chris Atkins, the coupon man; Tom Ness, the inventor who flapped his arms and croaked in the tops of redwoods. There was something sweet about this man who could be so kind with his strange friends, so passionate about trees. He seemed isolated in that horrible empty apartment, but he had made her laugh, he had called her weird, he had teased her about climbing barefoot. She had enjoyed teasing him back. She began thinking as she drove along, What am I going to do?

She arrived at the house in Corvallis on Sunday evening. Ted was sitting on the couch, eating a bowl of ice cream and watching television. When she started talking about her experiences in the redwoods, she thought that he was looking around her at the TV.

It clarified things.

Two days later, Antoine told Eldon that she was leaving him. He was taken completely by surprise. She sent an e-mail to Sillett telling him that she was planning to leave Eldon, but she didn't inform him of any plans of her own, other than to say that she would be traveling to Canada to be with her father.

Sillett e-mailed her back, asking if she could meet him somewhere for a cup of coffee. Antoine said that would be all right, and Sillett got into his truck and drove to Corvallis, a six-hour drive. They went for a walk in a park and talked about inconsequential things, as if he had just happened to drop by Corvallis. At some point, they stopped and found themselves looking at each other with nothing to say. He kissed her tenderly on the forehead, but he couldn't bring himself to kiss her on the lips. They left the park and got coffee.

"The expedition I'm planning to Australia next year," he said in the coffee shop. "I'm wondering if you'd like to join it as a climber." She said that she would like to join the expedition to Australia.

They began planning the expedition, going over details of financing and making a list of the equipment that she would need to bring. Their talk was strictly professional, but it was obvious to both of them that they were falling in love. At the end of their meeting, he paid for the coffee and drove six hours back home.

SHORTLY AFTER HE MET MARIE ANTOINE FOR COFFEE, SILLETT FLEW to Pennsylvania to visit his family. While he was there, he and Antoine communicated frequently by e-mail. Their relationship still seemed to be mostly in writing. Finally, he wrote this to her: "Let's get a room somewhere and stay in it." Sillett flew into Portland, where Antoine met his flight, and they ended up at the Radisson Hotel at the Portland airport. They spent thirty-six hours there, living mostly on room service, though they did get out for a short hike in the Columbia River gorge. "We were finally able to express our feelings for each other," Sillett said to me.

Then they had to part again. Antoine went to Canada to be with her father, and Sillett went climbing in the giant sequoia forest in the Sierra Nevada.

Ronald Antoine had a circle of friends at the lake, and they had been looking after him. When Marie arrived, she found that her father's illness had progressed rapidly. He refused her help, and claimed that he was feeling better, but he was almost unable to walk or take care of himself. She told her father that she was getting involved with a fellow canopy researcher. When Sillett called to speak with Marie, Ronald wanted to have a word with him. "So I gather you're a famous Ph.D., Steve?" he said. "Tell me about your research. Are you a tenured professor? Tell me about your interest in Marie. How do you feel about the kid? Can she be safe climbing these trees?"

Sillett reassured Ronald that Marie would be safe.

At the end of his visits to the island each summer, Marie's father always wrote down a list of groceries that he needed to buy the next summer. He drew up the grocery list just before he left this time, knowing full well that he would never see the island again. He barely managed to walk down the steps to the dock, the steps that he had built for Elizabeth.

STEVE SILLETT AND MARIE ANTOINE DROVE UP TO THE OLYMPIC Mountains and hiked into the rain forest by Lake Quinault, carrying climbing and camping gear. They piled it at the foot of Windigo. Sillett used a nylon cord that he had left in the tree to pull a rope up into it, and they hauled two Treeboats to the canopy. When they reached the top, they rested and talked before setting up their camp. They ate some dinner—bread, cheese, raisins, and peanuts—and sat near where Antoine had first learned how to stand on a branch.

They had never made love in the forest canopy, but they wanted to. The tree-climbing saddle was a problem, with its buckles and the equipment attached to it. The fact that there would be safety ropes hanging around and getting in the way would also hinder or prevent what they wanted to do. They talked it over and decided that the only way to do it was to remove all their ropes, safety gear, and clothing: to detach completely from the tree.

They strung up the two hammocks with one hammock positioned above the other. The lower hammock would be empty. It would, in theory, be available to catch anyone who fell out of the upper ham-mock. Anyone who fell would, in theory, be able to grab the safety hammock, hang from it, and climb or roll into it. They were both in top physical condition, and they felt that either of them could do this if things got out of control.

After they had gotten the hammocks rigged and positioned, they suspended themselves on their spider ropes from branches overhead and, still wearing their harnesses, began to undress in front of each other, taking off all of their clothing except their pants, which were underneath the harness. Antoine went into the hammock first, lower-ing herself into it. Then, while he waited, she unclipped her rope and took off her harness, slowly sliding herself out of it, and then she took off everything else, very carefully. He stayed close to the hammock and to her, so that if she wobbled or became unbalanced he could grab her and keep her from falling out. When she had stretched out in the Treeboat without anything on, he lay down in the hammock next to her. He unclipped his rope, detaching himself from the tree. He un-harnessed himself and undressed, lying next to her. It was a cloudless summer night, warm and almost windless, and the moon was half full. They made love for a long time. The aura of danger surrounding it—the need to stay in control as climbers in a situation where perfect control over one's body is impossible—added a sweet edge of hazard to their explorations.

▪ ▪ ▪

SOME TIME LATER, TIRED AND SPENT, THEY WERE HOLDING ON TO
each other in the hammock. He began to fall asleep. He rolled over
onto his back and crossed his arms over his chest. He opened his legs
wide and planted the soles of his feet together. One of his knees
flopped down over her leg, and his elbow dug into her ribs. He fell
asleep instantly.

She was lying on her back wondering if he was just dozing or if
this was it, if he was out for the night. He was crushing her. The hammock was too small for two people. She couldn't seem to fall asleep.
Her lover's frog-kneed position was untenable. "Steve?" She nudged
him. "Steve? Could you move your knee?"

He muttered something. He drew his knees up and then flopped
one of them back down over her legs.

"You're crushing me with your leg, Steve. I can't sleep. I think you
need to go down into the lower hammock." She gave him another little nudge.

He came half awake, sat up, and obeyed her. He got out of the
hammock, stepped out into the air, and lowered himself into the catch
hammock with his hands. At that moment, he woke up. My God,
what am I doing? he thought. This woman just kicked me out of bed!
I'm naked and wandering around at three hundred feet without a
rope.

"Sorry, Steve."

"No problem, Marie. Good night."

"Good night, Stevie."

NEWFOUND WORLD

ARIE ANTOINE MOVED IN WITH STEVE SILLETT IN AR-
cata. They split the rent on the apartment, and she
began seriously learning how to climb tall trees. There
were subtleties in the technique, highly specialized pieces of climbing
equipment to learn how to use, and climbing moves that are done
only in tree space and in no other vertical environment. She had to
learn how to use each piece of life-support gear correctly, a hundred
percent of the time, with no mistakes, because there is no such thing
as a short fall out of a redwood. She went slowly at first, slowly for a
long time, thinking about each move. On rainy days, in stormy
weather, she stayed at home and wrote, working on her master's thesis on Lobaria.

The Atlas Project, which Sillett had begun in 1996, was nearing
completion. The work of mapping and describing the Atlas Grove
had been far more than the work of one person. It involved Robert
Van Pelt, along with a number of undergraduates and graduate students, university faculty and staff, and Marie Antoine (who became a
member of the faculty at Humboldt State in 2003). The graduate stu-

dents and professional climbers included Mark G. Bailey, Billy Ellyson, James C. Spickler, Cameron Williams, Anthony R. Ambrose, Christine A. Ambrose, Gregory M. Jennings, Leah Larson, J. Brett Lovelace, Giacomo Renzullo, and others. Another important contributor to the work was an equipment technician named Douglas Ballantine, who died of a brain tumor in 1999. Sillett and his colleagues named one of the larger trees in the Atlas Grove Ballantine in his memory.

Here's what the canopy teams found.

The biggest tree in the grove is Ilúvatar, which stands near one end of the grove. It took Steve Sillett, three graduate students, and Marie Antoine, all of them highly skilled and in top athletic condition, roughly twenty days of climbing to make a 3-D map of the crown of Ilúvatar. One climber, usually Sillett, would take the measurements of a limb or a trunk using a measuring tape and a compass, calling out numbers to one or two other climbers, who wrote them down and performed other tasks. The numbers went into a large and growing database, an image of the Atlas Grove in a computer file, where the structure of the grove could be re-created in 3-D.

Ilúvatar contains 220 trunks. The crown of Ilúvatar fills thirty-one thousand cubic yards of space. "The top of Ilúvatar is so dense with foliage that you could put on a pair of snowshoes and walk around on top and play Frisbee there," Cameron Williams said to me. Sillett and Van Pelt performed a calculation that shows that Ilúvatar contains 37,500 cubic feet of wood.

They mapped the deep architecture of Ilúvatar. For their map, they made use of an insight that Francis Hallé had in his airship, the *Radeau des Cimes,* flying across forest canopies. Hallé was fascinated with the forms of trees. He wondered about their shapes, their patterns as they filled space. Trees are like snowflakes, in that no two trees in nature are alike. And yet, like snowflakes, trees are governed by rules as they grow. Even though all trees are different, the members of a given species of tree have a distinctive look that is often obvious at a glance and is unique to the particular species. An elm tree grows into a graceful wineglass shape, while an English yew has a billowing, squat architecture, with strong, flexible branches (suitable for carving into the English longbow). Many trees thrust out smaller "trees" from their limbs that look like versions of the main tree. These versions of the main tree thrust out still smaller versions. Hallé called the process reiteration. The tree reiterates itself, making smaller and smaller copies of itself as it attempts to fill space and gather light. This is the mathematical form called the fractal. A fractal is a shape that echoes its own shape at smaller and smaller scales of size. Not all trees reiterate themselves, but many do. Hallé's simple observation showed how a tree creates a complex structure in many scales—an architecture made up of nooks and crannies and shaded, moist spots and fertile pockets where all kinds of living things can become established and can interact with one another.

Ilúvatar had reiterated itself again and again, forming Ilúvatars within Ilúvatars, until it had become a kind of forest inside a forest. Sillett's map of Ilúvatar showed that its 220 trunks had been reiterated into six levels of hierarchy. Ilúvatar is one of the most structurally complicated living organisms that have ever been discovered.

THE FERN GARDENS IN THE ATLAS GROVE (AND IN THE GROVE OF Titans) are the second most massive collections of epiphytes—plants that grow on other plants—in any forest on earth. Robert Van Pelt discovered that the epiphytes in a rain forest on the Olympic Peninsula, in Washington, have an even greater mass than the epiphytes in the redwood forest. Both temperate rain forests have far larger and more massive collections of epiphytes than do any rain forests in the tropics. Some of the fern mats in the redwoods weigh two tons when they're saturated with water after a rain.

The canopy soil in the redwoods is teeming with soil mites—tiny arthropods of many different species. The canopy scientists have so far identified fifty-five different species of mites living in redwood canopy soil, and some of them are as yet unidentified or undescribed species. Some of the mites are primitive creatures called collembola, or springtails. There is such a variety of mites in the fern mats that Sillett began to think that redwood rain forests have a greater biodiversity of soil mites than do tropical rain forests.

A zoologist at Humboldt State, Michael A. Camann, climbed in the Atlas Grove and took samples of the fern mats and discovered

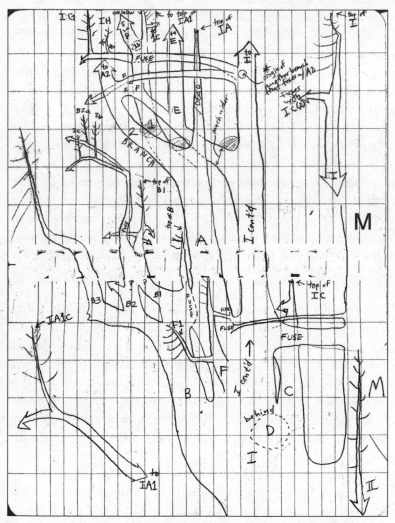

NOTES FROM ILÚVATAR. Two pages from Steve Sillett's climbing notebook, drawn in 1999, showing his developing map of one section of Ilúvatar's crown. This is a sketch of an eight-and-a-half-foot-thick trunk that gives rise to ninety-eight other trunks. It arises from Ilúvatar's main trunk (indicated by *M*) 126 feet above the ground. Several of the trunks and limbs have fused to create a lattice of hanging buttresses. It is probably a millennial structure—more than a thousand years old.

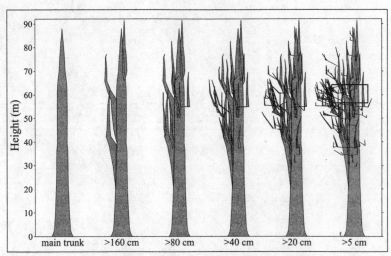

THE TRUNKS OF ILÚVATAR. The 220 reiterated trunks of Ilúvatar in six views, show-ing progressively smaller trunks in each view. This schematic map gives a sense of how a redwood adds trunks to itself over time. Note the tiny human figure on the right, for scale. The tree's limbs are shown as thin horizontal lines, but its branches are *not* shown (if they were, the map would be an impenetrable fuzz). The rectangle shows the approximate location of the illustration on the next page. From Stephen C. Sillett and Robert Van Pelt, "Structure of an Old-Growth Redwood Forest Canopy," *Ecological Monographs*.

WITHIN ILÚVATAR. A small part of the crown of Ilúvatar. Everything in this scene is part of Ilúvatar, except for the climbers (there are four of them) and some treetops in the lower right corner. This is the same general area indicated by the small box on page 205. *Drawing by Andrew Joslin, after a photograph by Thomas B. Dunklin.*

that they are also sprinkled with tiny aquatic creatures, crustaceans of an unnamed species of copepod. Copepods are oval-shaped, shrimp-like creatures, barely visible to the naked eye, that are sometimes called the insects of the ocean. They are the most abundant animals in the ocean, a huge constituent of plankton and a major part of the diet of baleen whales and of many small fish. There are some fifteen thousand known species of copepods. In the rain forests of Northern California, copepods live in the gravel beds of clear, running streams. Nobody knows how the "insects of the ocean" got into the redwood canopy, or how they carry on their life there. George Koch thought it was possible that the copepods may somehow be able to ascend redwoods by swimming up streams of water dribbling down the bark during rains. Sillett wondered if they got into the canopy by hitching rides on ospreys. Ospreys build nests in redwoods, and they feed on fish that they kill in the ocean, diving on the fish. They can't drink salt water, though, and so they depend on fresh water from streams. Sillett thought it was possible that ospreys landing in streams to drink might get the copepods trapped in their moist feathers and bring them into the canopy. When the birds landed in their nests, the copepods might fall off and start swimming and wandering in the fern mats, since the mats are full of water. They might feed on fungi there. It was just an idea. No one knows how the copepods make a living in the redwood canopy.

In 1995, a biologist named Hartwell H. Welsh found a nest and the eggs of a species of salamander called the wandering salamander in a hole in the trunk of a redwood that had just fallen down and was lying on the ground. The hole was sixty-five feet up the tree. Salamanders are amphibians, like frogs. The wandering salamander is brown, with gold spots, and like all salamanders is shy. It feeds on insects, mainly at night, and it lives in the cracks and holes in rotting wood. It occurs along the coast of Northern California. Biologists had always assumed that the wandering salamander lives only on the forest floor. The wandering salamander has no lungs and no gills—it absorbs oxygen directly through its skin. If its skin dries out, it quickly dies. One would not think it could survive in the canopy.

Some biologists wondered if Welsh's salamander had climbed into the hole and laid its eggs in the redwood *after* it had fallen down. As he was mapping Atlas Tree one day, Steve Sillett found a dead, mummified wandering salamander inside a knothole near the top of the tree, at 289 feet above the ground. In 2000, he found a live wandering salamander at 305 feet, living, evidently comfortably, in holes in rotten wood in the very top of Gaia Tree.

Jim Spickler decided to survey the Atlas Grove for wandering salamanders. He placed small salamander shelters in the soil on the branches of redwoods—little chambers made of screen and pieces of wood, set up in such a way that a salamander might like to hang out in them. He made rounds regularly, climbing into the trees and putting little tags on any salamanders he collected from the shelters. He found the same individual salamanders three years in a row in the same spots in the trees. The salamanders appear to be very territorial, and each one seems to claim its own piece of the forest in the sky. Spickler also found tiny young wandering salamanders, juveniles the size of a pin, living in the redwoods. These near-baby salamanders couldn't possibly have climbed up into the redwoods. The salamanders apparently breed in the canopy, perhaps living their entire life cycles in the air.

The fact that the wandering salamander needs to be moist all the time suggests that there is always water somewhere in the redwood-forest canopy. The trees hold huge amounts of water in their crowns. No one yet knows exactly how much water the redwood canopy can store, how the water moves through the canopy, or how, exactly, it's being used by the plants and animals that live there. The redwood-canopy teams are beginning to get some idea, though, of how much water may be aloft in the Atlas Grove. They think that in a hectare of redwood forest (two and a half acres) there may be between twenty-five thousand and fifty thousand liters of water in the structures in the trees. The researchers have, in effect, discovered aquifers aloft. They've run tests on the amount of water that some of the fern mats can hold, and they've found that, even in dry seasons and dry spells, the fern mats lose water very slowly. They recharge slowly, too, and it can take weeks for water falling from the sky to saturate the biggest fern mats.

As for the trees themselves, water that flows into a giant redwood through its roots takes two weeks or longer to reach the top of the

CANOPY PLANKTON. An unknown species of copepod collected in an aerial fern garden high in the Atlas Grove. *Photograph by Michael A. Camann.*

tree, moving slowly upward through the tree's sapwood. The sapwood lies under the bark and surrounds the tree's heartwood, its hard, heavy, rot-resistant, red-colored center. The sapwood of a redwood is moist and yellowish in color. It is mostly dead tissue, with some (important) living cells in it. The sapwood transports water through the tree, whereas the heartwood is entirely dead and doesn't transport water. All it does is support the tree, providing a strong superstructure. Underneath the bark lies a layer of actively dividing cells called the vascular cambium. The vascular cambium is entirely alive. As the cells of the cambium layer divide, they produce wood on the inside of the cambium layer and bark on the outside. The cambium is the expanding, active front of the tree's living self, reacting to influences as it grows. The wood it creates contains tissues called secondary xylem, which transport water through the tree. Outside the cambium, in the inner bark, lies the layer called the secondary phloem, which transports food—sugars, for example—which comes from the tree's leaves (its needles); the food is made in the leaves during photosynthesis, and the phloem delivers it to where it's needed.

The water-transport system of a redwood is a network of narrow, hollow cells called tracheids. This network is able to move water up from the ground and through a looming structure, thirty to thirty-eight stories tall, and do it continually, hour after hour, day after day, year after year, without failing. The system pulls water upward against tension. The pull originates in the leaves. As water evaporates from the leaves, a strong cohesive force between water molecules pulls more water upward through the network. This cohesive force is capillary action, and is the result of a phenomenon called hydrogen bonding: the hydrogen atoms in a water molecule tend to stick to nearby hydrogen atoms. In other words, water is sticky with itself and with other materials. This stickiness helps it move to the top of a tree.

The tree has to fight against gravity and against tension in the water. Water that's down in the ground wants to stick to soil particles, not go into the roots and up into the tree. The tension in the water in a tree is like the tension in a rubber band. Redwoods have an ability to stretch the rubber band very tight—they can suck water upward from the soil against almost overwhelming tension and gravity.

If the top of the tree dries out, or gets holes in it, tiny air bubbles, called embolisms, can appear in the network at the top of the tree. If enough air bubbles form in an area, the rubber band of water snaps at that spot and is permanently broken. A redwood can't restore the flow of water past air bubbles once they form, and everything above the air bubbles is doomed. The tree has had a stroke, and its top dies. A redwood can deal with a stroke. It simply grows a new top in a few centuries.

The cambium of a redwood, its ever-changing self beneath the bark, may be as thin as a single layer of living cells, invisible to the naked eye, identifiable only in a microscope. If the cambium of a giant redwood were spread out in a flat sheet, it would cover more than a soccer field, perhaps. A giant redwood that is adding one or two millimeters of thickness to its wood layer in a year is adding huge amounts of material to itself, and is one of the fastest-growing organisms in nature. That a redwood seems to be growing slowly is merely an illusion of human time.

A forest-canopy biologist at the University of California, Berkeley named Todd Dawson installed sensors in the tops of redwoods that grow around Santa Cruz, and in Sonoma and Humboldt counties. He and his colleagues discovered that a redwood that's bathed in fog can take moisture in through its needles and send the water downward into its small branches. Todd Dawson suspects, but so far hasn't been able to prove, that redwoods can also send water from their needles all the way downward into their trunks. In other words, redwoods can reverse the flow of water inside them when it suits their needs. This is one reason why a redwood can grow so tall—it doesn't have to depend entirely on water that it gathers from the ground and pulls up to its top. It can gather water from the air. Redwoods feed on the sky.

There are two species of huckleberries in the redwood canopy: the evergreen huckleberry and the red huckleberry. Both are closely related to the blueberry (the kind that's grown commercially), and they are also related to the blaeberry, a wild, sour blueberry that grows in the Highlands of Scotland.

There are very few birds in the redwood canopy. Redwoods produce poisons in their wood and needles that discourage insects from feeding on them, and consequently many species of birds that feed on insects go elsewhere to look for food. The climbers found bumblebees feeding around the huckleberry gardens, though, as well as various species of beetles living in rotten wood. They also found large, pink earthworms, of an undescribed species, living in the canopy soil. These soil worms live their lives in the canopy. No one knows how the soil worms get up into the redwoods in the first place.

Cameron Williams began collecting and identifying lichens in the redwood canopy. Williams grew up in Southern California, and he has a passion for skateboarding. He wears a skateboard helmet when he climbs in redwoods. "I've been skateboarding practically all my life," Williams explained to me. "But I can't do it forever. So I decided to study epiphytes in trees." Williams had identified 183 different canopy lichens in redwoods. They included about ten species of lichens that he couldn't attach names to. They may be undescribed species or species that are simply very hard to identify.

The nest of a red tree vole was discovered one day on a limb of the redwood named Ballantine. The red tree vole normally lives in Douglas-firs, not in redwoods. It is a canopy creature—it lives high up in the firs—and is an herbivore. It feeds on the needles of the firs, stripping off the middle parts of the soft, tender fir tips and bringing these needles back to its nest to eat them. The red tree vole is a keystone species of the canopy ecosystem in the Pacific Northwest, because it is the main prey of the spotted owl. The red tree voles that are living in redwoods are sleeping in them but are commuting across the limb system into Douglas-firs to forage for needles, and they bring the fir needles back into their redwood homes. Apparently, they like living in redwoods—where they are possibly less likely to become a meal for a spotted owl, since the owls may not be looking for them in the redwoods.

In addition to mosses, the researchers found vascular plants growing all over the redwood canopy. A vascular plant, unlike a moss, has a circulatory system of pipes to deliver water to its various parts. They found rhododendrons in bloom at 150 feet in the redwood canopy. They found a shrub called salal all over the place. On the ground, salal is a sprawling shrub, but in the gardens of the air

it forms interlocking branches, weblike in structure. They found cur-
rant bushes and elderberry bushes bearing fruit, and salmonberry
canes, which occasionally put out fruit, too.

They began to notice small trees of various species growing here
and there, in trunks and limbs and in crotches, hundreds of feet above
the forest floor. The trees are the bonsai of the redwood canopy.

The canopy bonsai include California laurel trees. One grows at
320 feet in the crown of a titan named the Redwood Creek Giant.
There are western hemlocks growing as bonsai, Douglas-fir bonsai,
and tan-oak bonsai, occurring in a sort of aerial microforest at the top
of the redwood forest. These are small trees, and for the most part
they aren't putting out seeds or reproducing, except for the bonsai
hemlocks, which do seem to be reproducing in the canopy. Sillett once
found an eight-foot-tall Sitka spruce growing on the limb of the Terex
Titan. They discovered tiny buckthorn trees growing near the tops of
redwoods. The species is the cascara buckthorn, which is native to
western North America and was used by the Indians as a purgative.
They found a buckthorn tree growing near the fern gardens in Atlas,
and they found another one growing near the top of Zeus, the tallest
member of the Atlas Grove. Climbing near the grave of Telperion one
day, Sillett came across a sort of hidden bowl in the sheared-off top of
a nameless redwood. There, at three hundred feet, he discovered a
bonsai redwood. Seven feet tall, it had rooted itself on a parapet in the
tree and was trying its best to dominate the canopy and cast its shade
on the monsters around it. It wasn't doing too well. Sillett didn't think
it would live very long.

A REDWOOD TREE IS BOTH MALE AND FEMALE—IT PRODUCES SPERM
cells and egg cells. The sperm cells are contained inside grains of
pollen. A redwood releases its pollen into the air from tiny male
cones, called strobili, which appear on the tips of branchlets. (Branchlets are the smallest branches, and they have needles on them.) Redwoods spew pollen from their male cones on sunny days in winter
and early spring. One day in January, Marie Antoine was climbing in
the Grove of Titans, near the top of the redwood named Sacajawea,
and the tree began to feel spring in the air and threw off so much

pollen that she began coughing and choking. The air was yellow from the tree's pollen. "Despite its name, Sacajawea seems to be more male than female," Antoine said.

A redwood's egg cells are in seeds in the female cones. The female cones hang from the branchlets, but they're larger than the male cones and have a knobby shape. Pollen gets into the cones—it may have drifted there from another redwood, or the redwood may be fertilizing its own seeds with its own pollen, depending on such factors as wind conditions. Inside the female cones, the grains of pollen split open, releasing sperm cells, which find their way into the seeds. A year later, the fertilized seeds fall out of the cones to the forest floor. The seeds of a redwood are tiny, and most of them don't grow. Like other conifers, redwoods produce most of their seeds only in certain years, when conditions are just right, which are called cone years. In a cone year, the trees produce large numbers of fertile cones—the branches can appear to be utterly studded with cones—which open up and drop vast numbers of seeds. Cone years seem to occur very infrequently in redwood forests. "We actually don't know how often redwoods have a cone year," Sillett said. "In all the years I've been climbing in redwoods, we've only had one good year for seeds, and that was in 2003. The rain of seeds was intense." The seeds dropped from the cones so profusely that they clogged the rain-collecting instruments that the scientists had installed in the trees. Over the course of its lifetime, a redwood may produce a billion seeds. On average, in the fullness of time, one of the seeds may grow up to become a mature redwood.

Redwoods are gymnosperms. The gymnosperms produce seeds and pollen, but they don't have flowers. Gymnosperms are distinct from the flowering plants, or angiosperms. Pine trees, hemlocks, and firs are gymnosperms, as are cedars and cypresses, and so are gingko trees (which have seeds but no flowers). The gymnosperms are older than the flowering plants; they arose earlier in the history of life on the earth.

The rise of the gymnosperms—their appearance and their diversification as life forms—occurred starting in the early Permian period, about 280 million years ago. Before that, in the Carboniferous period, between 360 and 280 million years ago, the earth was covered with forests made up of plants called Lycophytes, or, commonly, the lycopods. They were trees of a kind, but they looked like something from Mars, quite unlike the trees of today. The lycopod trees were tall and angular and bony-looking, with bark that resembled lizard skin or fish scales. Some of them grew to be 130 feet tall. (The lycopod trees and their spores, petrified in the form of coal, are now adding to global warming, as they are being burned in electrical power stations around the world.) The lycopod trees didn't have true leaves, but they had leafy structures that covered their trunks and limbs as well as sprouting from their branches. They also didn't have seeds, but they produced spores that came out of football-shaped spore-producing bodies on their branches. They certainly threw off vast amounts of spores, and the spores may have been extremely flammable in dry weather, enough to perhaps trigger bomblike, raging forest fires once in a while in lycopod forests (no one knows). Conifers (pines, spruces, firs, redwoods), being gymnosperms, make seeds. Seeds work rather well. In the endless *Iliad* of the trees, the lycopods lost their wars with the seed-bearing trees, and the lycopod forests were replaced by gymnosperm forests. Climbing in the Grove of Titans one day, Steve Sillett discovered huge, hanging draperies of a plant called Selaginalla. It is a lycopod. In effect, the lycopod forests ended up as decoration on redwoods.

The redwoods are very old as a kind of tree. Their origin in the systems of life on the planet occurred during the rise of the gymnosperms, sometime after 280 million years ago; no one is quite sure when. In 1985, a team of paleobotanists exploring a mountain in Antarctica found two fossilized seed cones, each the size of a jumbo olive. Because of their distinctive shape, the cones clearly came from some sort of redwoodlike tree. This ur-redwood lived in Triassic times, about 240 million years ago, when Antarctica was part of the supercontinent of Pangaea and was covered with temperate conifer forests. This was before the earliest known dinosaurs appeared on the earth.

Another fossilized redwood seed cone—one that looks almost identical to cones from modern redwoods—was found in France. This fossil cone dates from perhaps 190 million years ago. That was the early Jurassic period, when dinosaurs were all over the place. We can therefore hold in our mind's eye a nice image of dinosaurs wandering through redwoodlike forests in what is now France. But the fossils of ancestral redwoods that have come down to us from the age of the dinosaurs are only a collection in a shoebox, just bits and pieces of trees—a fossil branchlet here, a bit of fossil wood or bark there, fossil grains of pollen; we don't really know what the whole trees looked like.

Sixty-five million years ago, at the end of the age of dinosaurs, an asteroid or a comet, which was about six miles across and was traveling on a disturbed orbit, slammed into the earth in what is now the Yucatán Peninsula, in Mexico, causing an explosion of nature-altering power. This event, called the K-T impact, enveloped the earth in dust and rains of molten glass. Tsunamis a mile high roared over continents. The shock of the impact may have caused volcanoes to erupt. There may have been acid rains, worldwide forest fires, global darkness, global heat, or a global deep freeze. The K-T impact was most likely the cause of mass extinctions of many species of animals and plants all over the earth at that moment in time, including the dinosaurs.

The redwoods hardly seemed to notice the asteroid. If a redwood is burned in a fire to a blackened skeleton, it can come back to life, as long as its root system remains intact. The K-T impact had no evident long-lasting effect on the redwoods. It's possible that, after the impact, the redwoods sprouted up from the remains of their root systems, rising up in fairy rings in a ruined world—we don't know.

Redwoods flourished in the Eocene period, in the dawn of the age of mammals. During that time, some fifty million years ago, redwoods, or trees very much like them, existed in forests all over the Northern Hemisphere, from the polar sea to near the equator. Redwood fossils from early in the age of mammals have turned up on Axel Heiberg Island and on Spitsbergen Island—both are in the Arctic Sea—and in Siberia, as well as in Colombia, South America. This was the peak time of *Sequoia,* the high noon of the redwoods. Fossils of Eocene redwoods have been found in Wyoming, Nevada, and Idaho. Those North American redwoods looked almost exactly like the *Sequoia sempervirens* of today. They may even have been *Sequoia sempervirens*—the very same tree.

The human species, *Homo sapiens,* originated around two million years ago in eastern Africa. Modern man is thought to be about two hundred and fifty thousand years old. The modern redwood seems to be twenty million to perhaps as much as fifty million years old. In other words, *Sequoia sempervirens* could be eighty times older than modern *Homo sapiens.* As the age of mammals went along, the earth's climate cooled off and became drier; ice caps appeared on the earth, and the redwoods lost their wars with other trees and gradually retreated. They ended up occupying a few last redoubts scattered along what is now the coast of California. The coast redwood is a so-called relict species. It is a tiny remnant population of a life form that once spread in splendor and power across the face of nature. The redwood has settled down in California to live near the sea, the way many retired people do.

In the fall of 2001, Steve Sillett and Marie Antoine spent most of their waking hours in the air, moving on ropes up and down the trees in the Atlas Grove. Sillett needed to get the last measurements on each trunk of each tree in the Atlas Grove Transect, the area he had marked out for special study. There were literally thousands of measurements that still needed to be taken. The tiny two-and-a-half acre plot of redwoods contained thousands of aerial trunks, all connected to a handful of major trunks, the main shafts of the titans. The fall weather stayed calm, warm, and cloudless, but the winter rains were due. Hoping to get the work finished before the rains hit, they climbed in the Atlas Grove for up to twelve hours at a stretch, as long as there was any light.

One day in October, Antoine and Sillett collected the last bit of mapping data for the Atlas Project. The tallest member of the Atlas grove, Zeus, stands at the west end of the grove, beside the titans Rhea and Kronos. They climbed Zeus in the morning and took the final measurements of it. As evening came on, they rigged up a Treeboat between some trunks near the top, where there was a good view, near a bonsai buckthorn tree. They admired the buckthorn tree, and Antoine broke out a loaf of olive bread. They were each carrying an avocado in their climbing kits. They slit the avocados and squeezed the flesh onto the bread, and watched the sun go down over a ridge to the west.

"I think there's a chance, just a chance, that we can actually appreciate and understand how redwoods work in our lifetimes," he said to her. "I wouldn't have said that the first time I climbed here."

She squeezed some avocado on the bread and took a bite.

"Marie, do you think you could keep doing this for a while?" Sillett went on.

She swallowed. "Sure, I know I could keep doing it."

"Marie Antoine, will you marry me?"

She almost fell out of the hammock. It was a real surprise. She had assumed that they would be together for a long time, but for some reason this hadn't occurred to her. "Yes, Steve," she said. "Yes, I will."

As they finished eating the bread, they started talking about practical things. They would be traveling and climbing in foreign countries, in unexplored forests. They would be climbing wild trees in Australia. Could she get life insurance through his insurance policy? Medical insurance?

They thought they would let some people know. Cell phones sometimes work at the top of the redwood canopy, and they used one to call Ronald Antoine in Santa Barbara.

"Where are you, Marie? You're at the top of a redwood? What? You're engaged? That's great." He lowered his voice, and said, "Just run off with him, Marie. Just elope. Don't bother with all the wedding things."

THE PROBLEM WAS TO FIND A MINISTER WHO COULD CLIMB A REDwood. They engaged the services of a man of the cloth named Michael Furniss. He was a bearded geologist and an ordained minister of Sufi Universalism, who was willing to take a crack at it. He was licensed to perform a marriage ceremony valid in the State of California, so long as it took place on or above land in the state. Antoine and Sillett invited a handful of guests, and they scheduled their wedding for December 8, 2001.

As the wedding day approached, Antoine wondered what she would wear. Certainly a traditional gown wouldn't work with all the ropes. She went to a fabric shop in Arcata and, for five dollars, bought a piece of white material that was eight feet long. She made a veil, sewing pieces of Lobaria lichen into it to form a delicate pattern. It was as long as a wedding train, and would serve that function in tree space. The rest of her outfit would be rain pants and a sweater, with boots. The ceremony would occur in a rain forest—rain had to be expected, and it would enhance her veil, since lichens come to life when they're moist, and their colors brighten.

Soon after they became engaged, Sillett and Antoine climbed into the top of El Viejo del Norte, in the Grove of Titans, to do some work there. While he was busy with something on the ground, she waited at the top of the tree. The titan has a dead spire for a top—a silvery skeleton, sixty feet long, like the tower of a ruined cathedral. Antoine was suspended near the base of the dead spire, and she plucked a tiny sprig of foliage from a spray of the tree's highest living part. She took it to a jeweler and had it cast into two identical gold rings, each in the form of the sprig of redwood foliage.

For the wedding site, they selected a pair of giant trees that grow in a hidden spot in one of the valleys of the North Coast. The trees, called the East Spire and the West Spire, had been discovered by Michael Taylor and Steve Sillett during their wanderings. There is a sort of void that extends between them down to the forest floor. Sillett and Antoine decided to be married in midair, in the space between the spires.

The day before the wedding, they began rigging the ropes; they staged a sort of rehearsal between the spires. After they'd finished it, they prepared to rappel down to the ground. Sillett was in the West Spire, and Antoine was at the top of the East Spire. He descended first. Antoine was distracted and had been thinking about many things having to do with the wedding. Without realizing what she was doing, she did the thing that had haunted her from the moment she began climbing redwoods: she made a mistake. She placed a descender device on her rope, oriented the wrong way. This meant that, instead of grabbing the rope, the device would slide down freely: there would be no brakes. This would cause a fatal, free-fall plunge to the ground.

She didn't notice what she had done. After she had put the device on the rope, she clipped herself to it and leaned out into space, getting ready to jump off. She was looking down at the ground. "Something began screaming in my mind," she told me. She paused and examined the rappelling setup. It was all wrong. Lethally wrong. If she had jumped off, she would have fallen to her death out of the East Spire on the day before her wedding. A rush of vertigo flooded her. What am I doing up here? she thought.

Was it a warning? What was it? She was bathed in sweat. She sat down on a branch, wondering if she should tell Steve that she had almost killed herself. She put the device on the rope correctly and rappelled slowly down to the ground. She told him later. He was horrified.

ON THE MORNING OF THEIR WEDDING DAY, SILLETT AND ANTOINE climbed the opposite spires, and several witnesses climbed up and helped arrange the ropes. They included a bridesmaid named Becky Hix. The minister climbed up to a branch near the top, almost out of sight of the ground, and tied himself to it.

It was a sunny day, with light winds. Antoine and Sillett left their trees and moved toward each other in space, advancing along the rope strung between the spires. Her wedding veil, decorated with lungwort, floated around her in the wind. They met in the space between the trees and exchanged the rings, and they began to exchange private marriage vows.

Their plan, after giving each other their private vows, was to advance together along the rope to the minister standing on the branch, where they would give their public vows and he would perform the rest of the ceremony. They were now hanging on the middle of the rope, floating and bobbing in the breeze, and the bride's veil was streaming. They had discussed what would happen next. Sillett wrapped his legs around Antoine, and in that way he carried her across the threshold of air to the altar, pulling her along the rope.

THE WEDDING RECEPTION WAS HELD AT A HOUSE ON THE BEACH. Steve Sillett got drunk. Michael Taylor showed up with Conni Metcalf, and Sillett informed him that he and Marie Antoine were headed for Australia to climb some tall trees. He grabbed Michael and pounded him on the chest. "You will come to Australia," he said loudly. *"You will climb a tree."*

Shortly after the wedding, the newlyweds flew to Santa Barbara to visit Ronald Antoine. He had been unable to attend the ceremony, but he assured them that he was feeling better, though it was obvious to them that he was in terrible shape. Five minutes after they left, he called a taxi and went to the hospital emergency room.

Immediately after they returned home to Arcata, Marie called her father's apartment to let him know they had arrived home safely, and got no answer. This alarmed her, and she phoned a cousin who lived near her father and asked her to look into it. The cousin quickly called Marie back, telling her that her father was in the hospital and was apparently dying. Marie was now in an agony of worry about him. She phoned him in the hospital and told him that she would fly immediately back to Santa Barbara.

He said to her that he did not want her to come back. He was at peace, and didn't want her to see him die, and it was his last wish. "I want you to remember me alive," he said to her. She agonized about it, and wanted to disobey him, and berated herself for agonizing about it, and finally she decided that she had to honor his request. She and Steve scattered his ashes in the stand of white pines on Treaty Island where her mother's ashes lie, in sight of the balsam fir that she had first climbed as a little girl. It had not changed much.

STEVE SILLETT AND MARIE ANTOINE HONEYMOONED IN THE DANUM Valley of Malaysian Borneo, and on the Hume Plateau in the Great Dividing Range of Australia, where they climbed in some of the world's tallest forests. One day, to their amazement, Michael Taylor showed up on the Hume Plateau in a rented car. He had taken up Sillett's challenge. He said that he might just climb a *Eucalyptus regnans*. But then he lost his nerve and decided that he wouldn't. He stayed with Sillett and Antoine in a little bungalow they were using, and stood on the ground watching them up in the trees. One night over dinner, Antoine pulled out a piece of paper and a pen, and she wrote up a contract for Taylor to sign:

> I, Michael Taylor, agree that I will take advantage of this amazing opportunity to climb in the Australian forest canopy, and I will climb a *Eucalyptus regnans* tree, and I will enjoy it.

Michael W. Taylor

He signed the contract. The next day, Sillett and Taylor went out into the forest and found a wild *Eucalyptus regnans,* and Sillett shot an arrow into it and ascended it. He strung two ropes along the tree and came back down. Taylor got himself harnessed in a saddle and began climbing the rope very slowly. He wasn't feeling well. He got a short way above the ground and stopped. Sillett climbed up next to him and began talking with him to keep him calm. Taylor moved up the rope again, a little at a time. He got higher and higher.

It was a sunny, hot day, and partway up the tree the sun really began to bother Taylor's eyes. "I don't know if I can make it, Steve," he said. "I know I promised Marie I would climb a tree, but I just don't think I can." He was blinded by the light, and afraid.

Sillett put his arms around Taylor and held him. Then he pulled him over to the shady side of the tree. The two men swung back and forth on the rope, wrapped around each other, with Taylor clinging to Sillett. "Don't worry, Michael, I've got you," Sillett said. "You're safe."

Sillett asked Taylor if he wanted to climb some more. He kept his arms around him, holding him in the shade. Taylor said that he was thirsty, and was getting dehydrated and dizzy. Sillett told him to wait right there. He rappelled down to the ground, got a bottle of water, and climbed back with it. "You gave me a roast chicken once," he said to Taylor.

Eventually, Taylor reached a big branch, twenty-two stories up in the air, and as he sat on it he remembered a strange feeling he'd had when he looked at the fallen Telperion with Sillett and touched the tree's upper branches: an eerie feeling that he would end up in the air with Steve. I'm going to be okay, he thought. My friend is with me.

5

INTO
THE DEEP
CANOPY

THE GARAGE

I FIRST MET STEVE SILLETT IN JANUARY 2003, WHEN I WENT TO VISIT him in Arcata. It was a moist day, and storms had been blowing in from the Pacific. The town smelled of redwoods and the sea. Sillett was in his office on campus, a windowless room lit by fluorescent lights. The room did not have a lived-in look. He was stretched out on a chair in front of a computer, working on some data from the Atlas Project. He was wearing olive-green climbing pants, a pullover shirt, and mud-stained boots. His hands were broad and supple-looking—a tree climber's hands. He was then thirty-four years old.

Sillett tapped on a mouse and a 3-D image of the major trees in the Atlas Grove appeared on the screen. "These trees are awesome," he said. The trees looked something like witches' brooms standing on their handles. He tapped the mouse again, and the trees began to rotate, as if they were mounted on turntables. "This is Zeus, Rhea, and Kronos," he said, pointing to a cluster of slowly spinning redwoods at the far end of the grove. "We had a storm a couple of months ago. The top of Kronos broke off, but it didn't fall to the ground, and the broken top is now leaning against Rhea. This is perfect, because Rhea was the wife of Kronos, while in the myth Zeus defeated Kronos."

I asked him where the Atlas Grove was exactly.

He gave me a guarded look. "This is something you can't print."

Sillett was worried about recreational tree climbers. I had come to Arcata to write a story about him for *The New Yorker* magazine, and he didn't want me to blow the big trees' cover by telling amateurs where they were. Recreational climbers regarded Sillett as a near-legendary figure, and he regarded them with alarm. Climbing a tree without permission—ninja climbing—is an accepted part of the culture of recreational tree climbing.

"Not only are the redwoods sensitive to damage from climbing but the whole habitat of the redwood canopy is fragile," Sillett said to me that first day in his office. "If people start climbing around in it for recreational reasons, it will inevitably be damaged." A rope running through an aerial moss garden, a stray kick of a climber's boot, and centuries' worth of soil and plants could be knocked off a branch.

We talked about his research, and I took notes. Then he said politely that he was busy, and that he needed to move on to something else. He closed the view of Atlas Grove on his computer and stood up. "Do you have any other questions?"

"Yes. I was wondering if I could climb in the redwood canopy with you."

No reply. He gave me a cool stare. There was a pause while his eyes moved away from my face and traveled over my torso and my arms, and finally his gaze settled on my hands. "Are you a tree climber?"

"Yes, I am."

BY THE TIME I MET SILLETT, I HAD CLIMBED A DOZEN OR SO TREES near my house, which is in New Jersey. This didn't impress Sillett. He said that he was happy to have me walk around on the ground in the forest with him and his colleagues, but he wasn't about to let me go up anywhere on a rope, and he wasn't going to tell me how he did it, either. Sillett kept his climbing methods virtually classified, because he didn't want the recreational climbers to know how he was entering or moving in the redwoods. As a result, probably fewer than twenty people in the world knew and could perform the Sillett tall-tree climbing techniques. Besides, he said, it was dangerous up there.

"These trees are gnarly. There are places in the Atlas Grove where I can't justify the risk of letting *anyone* climb," Sillett said. The broken top of Kronos, for example, which was leaning into Rhea, had rendered the west end of the Atlas Grove off-limits. The top of Kronos was a monstrous widow-maker weighing many tons, suspended twenty-eight stories above the ground. It could cut loose at any moment and calve down through the grove, destroying everything in its path.

STEVE SILLETT AND MARIE ANTOINE LIVED IN A SMALL YELLOW HOUSE in the hills overlooking Arcata. They gave me their guest bedroom. The Sillett-Antoine household had become the nexus of tall-tree climbing in California. On any given evening, cars and pickup trucks, burdened with ropes and climbing gear, crowded the driveway. Canopy scientists and graduate students hung out in the living room and the kitchen, cooking meals and eating together. The clothing was Gore-Tex, and the conversation was trees, trees, trees.

For the next several days, I tagged along after Sillett and a graduate student named Anthony Ambrose while they climbed in Humboldt Redwoods State Park and at Prairie Creek. They were joined by George Koch, the tree physiologist at Northern Arizona University. They spent the time mainly installing sensors and other scientific equipment in the trees. I spoke to them occasionally by radio, standing at the base of a tree.

I returned to Arcata a few weeks later, and this time I brought a duffel bag of climbing gear with me. Sillett looked at my gear and explained that most of it wouldn't do me a lick of good in a redwood.

Sillett was planning to go climbing in a vast old tree, full of caves and with bonsai growing on it, that had been discovered months earlier by Christine Ambrose, Anthony Ambrose's wife. She named the tree Artemis, after the Greek goddess, the virgin huntress. "You guys have named too many trees around here after male gods," she explained to the others. After a certain amount of hesitation, Sillett agreed to let me climb up Artemis along a black tactical rope using mechanical ascenders. He wanted me to stay attached to the main rope at all times. He did not want me trying any fancy stuff, like moving around in a redwood on my own. I assured him I wouldn't do that, since I didn't know how. I climbed Artemis with Sillett, Marie Antoine, and Robert Van Pelt, staying stuck to the main rope like glue.

The next afternoon, Sillett and Antoine went to a party, leaving me alone in the house. I wandered into the kitchen and made some tea. The door to the garage was in the kitchen, and it was unlocked. Sillett and Antoine stored their tree-climbing gear in the garage. I went in, feeling uneasy and a little guilty. Heaps of rope were piled in the middle of the floor, sitting on tarps, left there to dry. There were metal shelves stacked with all sorts of things: pulleys, cord, duct tape, carabiners, soldering irons, climbing pouches, grapnel hooks, fly-fishing reels. Climbing saddles and helmets were hanging on the wall like battle gear. Several motion lanyards were hanging there, too— spider-rope rigs. The rigs consisted of spliced ropes, carabiners of certain types, and short lengths of spliced rope—split tails—tied in Blake's hitches. The spider rigs had the grimy look of hard, professional service. I took out a small notebook and began making notes.

"I guess that lanyard you're using is about fifty feet long, isn't it?" I said to Sillett later.

"Sixty feet," he answered promptly.

"Do you make those splices yourself?"

"No way."

"Who does the splicing for you?"

He mentioned the name of a firm that spliced the main part of the spider rig to his exact specifications.

BACK HOME IN NEW JERSEY, I CALLED UP THE FIRM AND ORDERED A sixty-foot piece of tree-climbing rope with very particular splices in it, four special carabiners, and two split tails. When they arrived, I assembled and tied them into a spider rig.

At first, I practiced with the rig while I was standing on the ground—I heaved the ends of the rig over branches in a maple tree in my front yard. Then, wearing a climbing saddle and a helmet, I raised myself into the air on the rig and got about six feet off the ground. In an ash tree that grows off to the side of my house, I ascended sixty feet by throwing alternate ends of the spider rope over successively higher anchor points—that is, I spidered my way to the top of the tree.

I extended my circle of climbing, and began making trips into the forest canopy on the hill above my house. The forest canopy on my hill extends from about fifty to about a hundred feet into the air. It is a temperate deciduous forest, composed of the crowns of sugar maples, red oaks, chestnut oaks, white oaks, hickories, ash trees, tulip poplars, and some tall, beautiful old beech trees, which grow above a clear stream called Freestone Brook. The forest canopy in that spot is trackless and wild. When I began climbing in it, it had never been visited by people, and parts of it had never been seen by human eyes— many trees have places in their crowns that can't be seen from the ground. The animals that live in the forest canopy of New Jersey don't always seem to know what humans are. One day when I was climbing a wild maple tree, a northern saw-whet owl flew out of a hole and perched on a branch a few feet away, staring at me with topaz eyes. Then, seemingly unconcerned, it flew back into its hole, inches from my face. The owl had never seen a human suspended on a rope in the air, and it may not have known what I was, or identified me as a threat.

Migrating songbirds sometimes landed in branches ten or fifteen feet away, much closer than the birds would approach a person standing on the ground. At times, flocks of birds would sweep through the canopy and divide around me or move beneath me. During climbs into taller trees, I was occasionally able to look down on the backs of birds, which shine with reflected sunlight as they move through the green depths of the canopy, like schools of fish.

The canopy above Freestone Brook is populated with northern flying squirrels. The northern flying squirrel is a nocturnal animal, and is rarely seen by people. It is smaller than the common American gray squirrel, and it has silky brown fur and large, liquid brown eyes, and is a delicately beautiful, gentle creature. The flying squirrel has a membrane between its forelegs and its rear legs, which it uses as wings, for gliding. Flying squirrels exhibit calm, un-squirrel-like behavior. Sometimes I encountered one of them clinging to a branch, motionless and staring, placid and apparently unafraid. At night, they glide around in the trees looking for insects, and they land on the ground, where they dig for underground fungi—wild truffles—growing among the roots of trees. Flying squirrels have a keen sense of smell, and they are inordinately fond of truffles.

One day in the top of an oak tree, high above a pool in Freestone Brook, I came across three flying squirrels clinging to a branch. They were looking at me. Two of them moved away, climbing higher in the tree. The third continued to stare. I looped my spider rope over a branch and skywalked over to the squirrel. It still didn't move, just continued to stare. I reached out a finger and began to stroke it lightly between its ears, the way you would stroke a cat. The fur was extremely soft. The animal closed its eyes. But I must have moved, because after a few moments its eyes flew open, and it let go of the branch and plummeted down, caught the air, and soared away.

THE FIRE CAVES
OF ADVENTURE

I N THE DARKNESS BEFORE DAWN ON A COLD NOVEMBER MORNING in Arcata, Marie Antoine was hurrying around her kitchen, singing to herself in a dreamy kind of way. She wore a gray cashmere hoodie sweater, cream-colored slacks, and beat-up climbing boots. She tossed a handful of blueberries into a blender. "Steve, do you want a smoothie?"

"Definitely."

Steve Sillett and I were kneeling on the living-room floor, where he was inspecting a heap of my climbing equipment. He had said that I could climb again, and now he wanted to look at my gear. He picked up my climbing saddle and hefted it in a finicky way, like a restaurateur judging a flounder at a fish market.

Marie Antoine handed him a smoothie. She went back to the kitchen, and the blender ran again.

Sillett put my things down and went into the garage and got a sixty-foot length of bare rope, the backbone of a spider rig. "I'll make up one of these for you."

"You don't need to," I said.

I placed the rope on the living-room floor and took two split tails from my duffel bag. I put them on the floor in the proper way and tied the entire spider-rope system. I dressed and set the knots.

Sillett sipped his smoothie. "Dude, you're doing it." He picked the rig up and inspected it. "Where did you learn this?"

"In your garage."

"No way."

"I've been practicing with it a little."

"Sweet," he said, and handed the rig back to me. "You can stuff it in your climbing bag." A motion lanyard weighs eight pounds. When it isn't in use, it's normally kept in a bag clipped to the climber's foot stirrup.

Antoine came over and handed me a smoothie. "Which tree are we going to?" she asked.

"Richard's got himself a lanyard," Sillett said. "I think we need to go to Adventure."

ADVENTURE IS ONE OF THE TREES THAT MICHAEL TAYLOR DISCOV-ered in Prairie Creek Redwoods State Park around the time that he found the Atlas Grove. It is a lone titan; it isn't part of a grove. Adventure Tree is 334 feet tall, and it contains 31,000 cubic feet of wood. The main trunk is sixteen feet in diameter near ground level, and it maintains its huge girth nearly all the way to the top of the tree. The tree has a total of forty extra trunks. Adventure also has four fire caves in its crown—caves hollowed out by forest fires.

"Adventure Tree is never exactly my first choice for a tree to climb," Antoine commented, as she got her stuff together in the garage. "My first experience climbing that tree was kind of scary."

I asked her what had been scary about Adventure.

"I got lost in it."

WE WENT ALONG A TRAIL, WEARING BACKPACKS FULL OF ROPE AND gear. Antoine led the way, moving quickly. She weighed 120 pounds, and even with sixty pounds of gear on her back she still often walked faster than anyone else in a climbing team. She turned off the trail and began bushwhacking through steep underbrush. We moved in zigzags around masses of sword ferns. The ferns were chest-high and soaking wet; it had rained during the night. The trunks of redwoods soared into remote crowns. Blades of sunlight angled through the canopy, glittering with droplets of water falling from branches. The ground sparkled with redwood sorrel, and the sky was pale blue, mutable with running clouds.

We arrived at a small creek. A redwood log spanned it, forming a natural bridge. We crossed on the log. Adventure grew out of the bank on the far side: I saw a megacylinder of wood with a thermonuclear crown.

At the base of the tree, Sillett pulled the end of a black climbing rope from his pack and threaded it up through an anchor high in the tree and back down to the ground. While he was getting ready to climb, Antoine led me around to the other side of the tree. Starting at about a hundred feet above the ground, the trunk widened—Adventure Tree got even bigger than it was near its base, and out of the widened trunk emerged a towering system of extra trunks, some living and some dead, that ran upward along the stream side of Adventure for more than twenty stories. Antoine put her hands into her pockets and looked up. "The first time I climbed this tree with Steve, he told me to go to a certain place and I misunderstood him," she said. She ended up wandering among columns of rotten wood, which wobbled and seemed ready to collapse. Finally, she tied herself to a branch and called her husband on the radio and told him that she was lost. It took him twenty minutes to find her.

Sillett clamped a pair of ascenders to the black rope and began to jug up the trunk. A raven called somewhere in the upper canopy. This was followed by a delicate *pip, pip, pip* closer to the ground. It was a kinglet, a winter bird of the lower forest, which feeds on insects it finds in small trees. Sillett vanished in Adventure. High up in the tree, he rearranged the rope so that the two ends hung down along opposite sides of the tree. He used the radio to tell us that the ropes were ready.

We put on our helmets and climbing saddles. I clamped ascenders to my length of the black rope and clipped the bag containing the spider rope to my foot stirrup. I sat back in the saddle, putting my

weight on the rope, and began to jug upward. The tree was a furrowed wall of wood, coated with a foam of liverworts. Liverworts are
tiny plants that do not have a vascular system—no pipes. They originated back in the earliest days of plants. Twenty feet above the
ground, the tree's bark was pitted with fire scars. A small fire on the
ground had made the scars, perhaps within the past few hundred
years. Marie Antoine was climbing somewhere around the horizon of
the trunk, out of sight. I had no idea where she was.

I kept on jugging. Seventy feet above the ground, I passed a burl—
a benign growth in the wood of the tree. As I climbed, I touched my
feet lightly against the tree now and again. At ninety feet, there were
tent spiderwebs scattered all over pockets and cracks in the bark. I
wondered what the spiders were eating at this height. A moth fluttered
by. Perhaps it was a meal for a spider. I started swinging around the
trunk and encountered Marie Antoine. She had been circling toward
me to see how I was doing.

The bark of the tree was covered with a lumpy white crust that
looked like sugar frosting. "What is this stuff?" I asked her.

"Pertusaria. It's a lichen."

Pertusaria is also called wart lichen. The warts were mingled with
splotches of a grayish-green dust that was stuck to the bark. The dust
was a lichen called Lepraria. Supposedly, it looks like the infected skin
of a leper. The leprosy lichen was mixed with fingering spurts of the
lichen called Cladonia. There are many kinds of Cladonia, and they
are among the most beautiful lichens. They come in all kinds of
shapes—trumpets, claws, crowns, pinto beans on stalks, bones,
clouds, castles, javelins, blobs of whipped cream, and red-capped
British soldiers. This one looked like pale-green tongues of fire.
Around twenty different Cladonias have been found in the redwood
canopy, and even experts have trouble telling them apart. This one
seems to have been *Cladonia squamosa,* sometimes known as dragon
fire. Dragon fire is widespread in North America, and it occurs across
Europe from Italy to Scandinavia, and across Russia to the mountains
of China. Dragon fire has an uncommon beauty, and its presence in
the redwood forest reminds us that the redwoods exist in a tapestry of
forest ecosystems woven across the Northern Hemisphere. Scattered
near the dragon fire were clusters of orange disks that looked like tiny

pumpkin pies. Pumpkin-pie lichen occurs only in the rain forests of the Pacific Northwest. One certain kind of redwood-forest lichen doesn't occur on Adventure. It is a sickly yellowish-green, pinkish, splatty-looking stuff called fairy puke. The botanists haven't noticed it on any redwoods except the Terex Titan, which is one of the oldest individuals in the redwood forest. Perhaps Adventure hasn't lived long enough to have gotten fairy puke on it.

As I dangled on the rope, I put my face closer to the micro-landscape. There were yet more lichens growing in the cracks of the bark, lichens barely visible to the eye. The cracks were lined with black dots standing up on stalks, like the heads of pins shoved into the wood. Their common name is pin lichen, or stubble lichen (they look like an unshaved face). It occurred to me that in order to see a giant tree you need a magnifying glass.

I WANTED TO SEE THE ARRAY OF TRUNKS THAT LOOMED OVER THE creek, the dark side of Adventure, and I kicked off and traveled in that direction, getting closer to the goal in a series of gentle swings. I found myself in the middle of a Gothic tower of fusions, bridges, and spires, held up by flying buttresses. The zone was crisscrossed by branches, and the trunks ran out of sight in both directions, upward and downward. Overhead there was nothing but canopy. No sky, although when I looked down I could see a small patch of ground, starred with ferns. In front of me, at a height of 180 feet, was a fire cave. It was called the Upper Fire Cave, and its mouth was plastered with dirt. The dirt was canopy soil.

By gripping ridges in the bark, I was able to pull myself up to the lip of the cave, and I looked into it. It proved to be a sort of airy chamber in the underside of a flying buttress, and it opened downward into empty space. It was more like a fire ceiling than a fire cave. I ended up hanging in midair, a few feet below the charred ceiling, looking straight down at the stream. There was a faint sound of rushing water. Strands of computer cables emerged from the cave walls, where the canopy scientists had implanted electronic probes. I touched the wall of the cave. It was moist, and it had a yellowish color and a musty smell. It felt like Stilton cheese.

Two hundred and fifty feet up, the light became brighter, although I still couldn't see the sky, and the crown of Adventure billowed into a riot of living branches. By then, the ground had disappeared completely, hidden below decks of foliage in the lower parts of the canopy. This was the deep canopy—a world between the ground and the sky, an intermediary realm, neither fully solid nor purely air, an ever-changing scaffold joining heaven and earth, ruled by the forces of gravity, wind, fire, and time. At this height, the main trunk of Adventure was seven feet in diameter, and the huckleberry thickets had begun in earnest. They were evergreen huckleberries. In November, in the California rain forest, huckleberry leaves are tinged with scarlet at their edges. The bushes were all over the tree: perched on its branches, occupying its crotches, and popping straight out of cracks and holes in the trunk. I wormed my way through the bushes, following the black rope upward.

At 290 feet, I encountered Steve Sillett. He was sitting on a branch inside a spray of huckleberry bushes, and he had a thoughtful look on his face. The main trunk of Adventure had split open near where he sat, spewing out a riot of huckleberries and ferns, and revealing dead and rotten wood inside the tree. "This beast is full of rot pockets," he said. "These huckleberry bushes are putting their roots into the center of the tree. One summer we had half the normal rainfall, but these bushes still put out a full crop of huckleberries. They're getting their water from rotten wood inside the tree." He pointed to something on the side of the tree. "Check out that little brown moss over there."

"Which moss?"

"The one that looks like it's dead."

I wormed through the branches until I could see a greasy-looking thatch growing below a wound at the base of a stob. Redwood sap—a reddish lemony-smelling liquid—had been dribbling over the moss.

"It's called *Orthodontium gracile*," he said. It is also known as slender straight-toothed moss. "It's extremely rare. We've noticed it growing below wounds in old-growth redwoods. It seems to like the resin, and maybe it gets some nutrients from it, I don't know. It's nearly gone in Oregon, I think."

Antoine had come over to our spot, and she admired the grubby little moss. I had arrived at the upper end of the black rope. Nearby, I saw the bottom end of a second climbing rope, a white one, which was hanging down along the trunk of the tree. It led toward the top of Adventure, wandering out of sight. Sillett suggested that I go up.

I transferred my ascenders to the white rope and climbed up it, wriggling through a jumble of redwood branches and huckleberry shrubs. The bag that held my spider rope bumped along through the bushes. Abruptly, the crown thinned out and a view opened across Adventure Valley.

The rope came to an end about fifteen feet below the top of the tree. No ropes led to the top. In order to get there, I would have to detach from the main rope and use my spider rope. I took it out of the bag, attached it to my saddle with carabiners, and threw one end over a branch above me. I pulled the rope back to me, to form a noose over the branch, and clipped the noose to my saddle. Then I detached myself from the main rope.

There is something unnerving about leaving the main rope behind and going into motion in the crown of a redwood. The main rope is a lifeline that connects a climber to the ground, an escape route out of the tree. Once you disconnect from the main rope, you're on your own.

With my weight on the spider rope, I leaned back until my body was horizontal and my feet were planted on the trunk, and I walked up Adventure, cinching up the noose, making it shorter. I threw one end of the spider rope over a higher branch, clipped it back to my saddle, and trunk-walked to the top. I ended up hanging at the top of the tree in the middle of a bush studded with huckleberries. I ate a few of them.

The branches in the tree's top were festooned with frizzy lichens called witch's hair. There were beard lichens, including long, yellow-green hanging strands and loops of *Usnea longissima,* or Methuselah's beard, the longest lichen in the world. It can grow up to ten feet long. The early settlers in Oregon found Methuselah's beard hanging everywhere in the forests, and they filled mattresses with it and used it as a lining for babies' diapers. Today, it is rare.

The uttermost top of Adventure is dead. It is a gray trunk, encrusted with lichens, extending about six feet above the huckleberry bushes. The trunk ends at a sheared-off stump. Centuries ago, perhaps roughly around the time John Milton wrote *Paradise Lost,* the top of Adventure was blown off in a storm.

The branches around me trembled. A spider rope flipped over a nearby branch, and Marie Antoine appeared. She trunk-walked up to a platform of branches and sat on them. "The top of this tree is just a big old juicy dead-wood pit," she commented.

The dead trunk at the top of Adventure is a natural water tank, Antoine explained. Rainwater collects in the broken stump and runs down inside Adventure, where it saturates the rotten wood like a sponge. All kinds of plants send their roots into the dead wood, and it nourishes lichens and mosses as well.

There was something called tube lichen, or *Hypogymnia,* and there was *Platismatia glauca* all over the place near the top of Adventure— ragbag lichen, the species that Steve Sillett had discovered on Nameless as a college student. "There's also a lot of *Mycoblastus sanguinarius"*—black bloody heart—Antoine said, moving her finger around on the wood. The black bloody heart looked like dots of tar stuck to the wood, and the dots were smaller than this o. "When you pick one of the dots apart, you see that it's got bloody spots inside it," she said.

I split one with my fingernail and saw a gleam of brilliant red. Its heart really looked like a droplet of blood.

Marie Antoine sat back on the platform of branches, getting herself into the light, for warmth, it seemed.

"What drives you to climb these trees?" I asked her.

"When I first saw the redwoods in the Grove of Titans, I remember feeling a deep physical attraction to them," she answered. "When I'm climbing in the redwoods, I have a feeling of being one-on-one with the tree. The trees are something we can experience without any burden. I feel ultimately fulfilled as a person when I'm working in these trees—working to answer questions about them. It doesn't have to be me asking the questions about the trees, as I did with my Lobaria study—it can also be me working on the questions that others have asked but which are important. And I can begin asking questions of my own."

"Why is it important to ask questions about redwoods?"

"It helps us know how the forests work as a whole and how the trees work as organisms," she said. "Then we can help them out if they're having problems—and they *are* having problems. It occurs to me that I have a fairly cynical outlook on so many things in the world today—this insane world. But as long as we still have these trees, there's hope for us."

It was a sunny day, and a breeze was blowing, bringing a smell of salt air. Adventure rocked in the breeze, like a ship riding at anchor. A seagull floated by.

BEYOND THE REDWOODS

I CONTINUED TO EXPLORE THE CANOPY OF NEW JERSEY. MY WIFE and I have three children, who then ranged in age from eleven to fifteen. As they watched me climb, they began bugging me to take them up into the trees, and so I returned to the Tree Climbers International school in Atlanta for advanced training. I learned how to adapt the arborist climbing technique for children and untrained adults, to keep them safe. There are special ways to rig ropes in a tree, and special pieces of gear, such as child-size tree-climbing saddles.

I began taking my children up into the forest canopy. They proved to be adept as climbers, and they didn't seem to have the sort of fear of heights that adults often have. Our oldest child, Marguerite, said that she wanted to eat lunch in the top of a tree, so we climbed sixty feet into a white ash and had soup and sandwiches there. There was a big tree at the top of our hill, a red oak a hundred feet tall and more than a yard across near the base. I removed the dead branches from it, to make it safe, and named it the Red Queen.

One day in spring, I took our children Oliver and Laura camping in the Red Queen. I strung up three Treeboats at different heights, fifty to seventy feet above the ground, creating a campsite in the crown of the tree. We ascended the tree at sunset and ate dinner— hoagies and chocolate milk. We fell asleep in our Treeboats, anchored to the tree with ropes. The next morning, when we got to the ground, we discovered that we were a little unsteady on our feet, like sailors stepping off a sailing vessel. We had been suspended in the canopy for thirteen hours, and the Red Queen had been moving the whole time. Trees are always in motion, even in the calmest air.

In October, Oliver and I camped by ourselves in the top of a tree that grew above a pool in Freestone Brook. We were eighty feet over the water. In the gray light before dawn, he woke me up. "Look at this," he said.

He was sitting up in his hammock, and two flying squirrels were crawling over his sleeping bag. A third squirrel came out of his sleeping bag and perched on his head. Then, as silently as they had arrived, the flying squirrels leaped and fell off him, gliding away. Perhaps they had been exploring him for truffles. "I don't think the kids in school are really going to believe this," he said.

Laura said she wanted to learn more, so I took her to the tree-climbing school, where she learned how to skywalk and, at thirteen, became the youngest certified tree climber in the history of the sport. With the instructor Tim Kovar, we climbed a giant tulip poplar tree in the mountains of north Georgia that has a cave inside it. The mouth of the tree cave opens ninety feet above the ground. Laura climbed in through the mouth and rappelled down twenty feet through the center of the tree. She came out into a room inside the tree, where a hole looked out into the canopy, like a round window. "I kind of thought it needed a bell and a sign that said THE WOLERY," she remarked (referring to Owl's house in *Winnie-the-Pooh*).

My wife, Michelle, professed a fear of heights, but one day I persuaded her to climb up over Freestone Brook with me, and we rested in a Treeboat and chatted as the sun went down. After she had rappelled to the ground, she said that she wouldn't mind doing it again. Mysteriously, almost unaccountably, my family had ended up in the trees, sort of like the Swiss Family Robinson.

EXPLORATION. The author's thirteen-year-old daughter skywalking. She is using a simplified version of the spider rig used by redwood climbers. *Drawing by Andrew Joslin.*

■ ■ ■

For a summer trip, we went to the Highlands of Scotland to climb in the ancient Caledonian forest canopy. The Scottish Highlands are virtually devoid of trees. Two thousand years ago, 90 percent of the Highlands was covered with forests, but logging, sheep grazing, and grazing by red deer have almost eliminated the trees entirely. A few fragments of the old Caledonian forest have survived in the western Highlands, in and around Glen Affric, a little-visited glen that wanders between Loch Ness and the Isle of Skye. The old trees grow in sheltered pockets around lakes, and in rocky, trailless gorges that are jammed with boulders and heather, and are difficult to pass through on foot. I think it's very likely that we were the first tourists ever to visit Scotland in order to climb trees.

Glen Affric is filled with lakes and surrounded by mountain peaks that resemble worn molars. The trees there are Caledonian pines—Scotch pines. The species name is *Pinus sylvestris*. The oldest individuals may be more than four hundred years old, but more typically they live to around three hundred, when they get scraggly and die naturally of old age. An old Caledonian pine develops a wide, billowing, mushroom-shaped crown that can be as much as fifty feet across—the trees are as wide as they are tall. As far as I could tell, the ancient Caledonian forest canopy had never been explored. No canopy researchers or biologists had ever gone there. Nobody, including Steve Sillett, had any notion of what we might find. It seemed almost incredible that any place in the British Isles could remain unexplored, or that, on a family trip with children, we would be the first humans to visit it. "Let me know what you see," Sillett said before we left.

We stayed at a B&B in Glen Affric, and the owner let us store our ropes and gear in his barn. Glen Affric is a remote place, favored mostly by backpackers and mountain bikers. With a little crossbow, I fired a fishing line into a monstrous Caledonian pine that grew in a heather-filled hollow near Loch Affric, in the center of the glen. I set a rope in the tree and climbed into it, pulling more ropes with me, and I rigged the ropes over anchor points, and eventually my three children climbed up. We were able to see the shape of the tree clearly from inside the crown. The crown formed a tangle of arched, wandering limbs and small platforms. The limbs were old and fat, and had become fused, flowing together in places to create flying buttresses. The architecture of the tree reminded me of a city drawn by Dr. Seuss. The tree seemed to be about three hundred years old, and it was wild: no humans had ever entered its crown, that was clear. When the children got up into the tree, they delighted in moving around on the ropes, traveling on diagonals through the air, poking around, branch-walking. I asked them not to walk on anything that was growing on the tree and to watch where they put their feet, and to remember that anything they knocked off a branch might take decades to grow back.

Our first big discovery was the existence of rowan trees growing as epiphytes all over the branches of the Caledonian pine, almost entirely out of sight of the ground. It was a bonsai forest hidden in the canopy. Rowan trees have red berries. Turning and twisting on their ropes, the children counted at least thirteen bonsai rowans growing on the branches of the Caledonian pine that we were climbing in. One of the rowans had berries hanging from a few of its tiny branches. This meant that the rowan trees were setting fruit and reproducing themselves in the canopy. The children also found a small blaeberry bush growing from a hole in a branch. The species name of the blaeberry is *Vaccinium myrtillus*. It is very closely related to the two species of huckleberries that live in the redwood forest canopy.

There were all sorts of mosses that I couldn't identify. The branches were covered with tube lichen, or *Hypogymnia*—the same lichen that I had seen near the top of Adventure Tree in California. Looking more closely, I found speckles of tar all over the tree. They looked like black bloody heart lichen, and seemed to be the same organism that Marie Antoine had pointed out to me. I split one of the black dots with my thumbnail and, sure enough, it had a crimson center, like a droplet of blood. There was ragbag lichen—*Platismatia glauca*—plastered all over the branches.

Laura skywalked through the tree, and turned herself nearly upside down in the air so that she could inspect a tiny aerial garden growing on top of a platform-like limb. The garden was no bigger than a salad bowl, yet it contained six rowan trees, sprouting from a bed of canopy soil, moss, and ragbag lichen. The old Caledonian forest began to disappear when the ancient peoples of Scotland established farms in wooded valleys, cutting down the timber. The final blow to the forest happened with the disappearance of the wolf. The last wolf was killed in Scotland in 1746, the same year that Bonnie Prince Charlie lost the Battle of Culloden and fled, they say, through Glen Affric to the Isle of Skye, while the British took over Scotland. With the wolves gone, the population of red deer exploded.

The Scottish red deer are, in fact, a type of elk. The wolf was the predator that kept their population in check. The deer were protected for hunting, and sheep were introduced all over the Highlands, too. Human hunters couldn't take the place of the wolves, however, and had little effect on the red deer, which overran the Highlands, crowding into stands of Caledonian pines in the winter to get out of the wind. The deer (even more than the sheep) grazed on the pine seedlings, eating every one. The Caledonian pines stopped reproducing themselves around the middle of the eighteenth century. Each year, the pines of Glen Affric drop seeds, and the seeds sprout and are eaten by the deer.

The pines in Glen Affric are all mostly around 250 years old, and some of them (including perhaps the one we were climbing in) began growing while wolves still roamed the Highlands. As we looked around, we could see no younger pines growing underneath them. We gazed down on a blank forest floor that was covered with just about nothing but ferns and heather—no young trees of any species. The Caledonian forest was doomed to vanish as the old trees died, because there were no trees coming in to replace them. During the twenty-first century, the last fragments of the old Caledonian forest and all that it contains will begin to disappear as the trees die of old age. This was a striking example of a total crash of a forest ecosystem evidently brought about by the removal of the top predator. The extinction of the wolf in Scotland would cause not only the loss of the pines but the loss of the lichens, the bonsai rowans, the smaller trees, and the animals and birds that depended on the trees—the extinction of the forest itself.

A dirt road ran near the tree, and a group of German tourists clattered by on mountain bikes. They didn't seem to notice us. "It seems strange that Bonnie Prince Charlie could have walked along that road," Laura said, suspended in the air near the rowan garden. "He could have looked at this tree."

A Scottish conservation group called Trees for Life has begun fencing off patches of the forest in order keep the deer out, and planting seedlings. Inside the fenced areas, the trees are reproducing naturally—we could see younger pines thrusting up among their elders. In the fenced slivers of Glen Affric, around A.D. 2250, if the trees are continually protected from deer, another old Caledonian forest will gradually rise.

"I felt as if we were seeing a version of the redwood canopy, except that it was in Scotland," I said to Sillett afterward. "Why, when you get up in the canopy, does Scotland look like California? What's going on?"

"This is awesome," he replied. "Nobody knew this. What you saw, I surmise, is a fragment of a truly ancient temperate-forest ecosystem, which stretched across the Northern Hemisphere millions of years ago, when Scotland was still joined to North America, before the continents pulled apart. Did you collect any samples?"

"No. I didn't want to remove anything from those trees," I said.

"Oh, man—one of these days we're going to have to get an expedition together and do the descriptive natural history of the Caledonian canopy. Nobody's done it."

ONE OF THE CONSEQUENCES OF MY INTEREST IN CLIMBING TREES WAS that I became a gearhead. Sillett and Antoine have a lot of great gadgets. One of them is a grapnel hook that looks and functions a little like the hook that Batman uses to climb buildings and snag things. The hook is attached to a fishing line on a fly-fishing reel mounted on the climber's saddle. The grapnel hook is used to catch and retrieve ropes and lines that are draped on out-of-reach branches. It is a solution to the common problem of how to get the end of a rope that's over a branch back to you so that you can form a loop.

The type of fly-fishing reel the redwood climbers use is an old-fashioned one called the automatic-rewind reel. They were popular from the 1930s to the 1950s: you push the button on the reel and the line reels in automatically. Fly fishermen now think these reels are tacky, and nobody wants to be seen on a river in Montana using a push-button fly-reel. You can get one for about five dollars on eBay.

I saw that Sillett had fitted his climbing harness with a Shakespeare Silent Automatic fly-reel dating from around 1950. It was forest-green and had an aerodynamic shape, like a fifties Chevrolet. I began to covet his reel. I started looking on eBay and eventually found a Shakespeare Silent, forest green, with an aerodynamic shape, in mint condition. It was perfect. There was a bid of three dollars on it. I bid four dollars. There was only one other bidder in the auction, called Gnarlmaker. Gnarlmaker was aggressive, and came back with a bid of six dollars. I waited, and at the last moment I put in a blowout bid of eleven dollars and got the reel.

Not long afterward, I was talking with Sillett and he began complaining about deteriorating conditions on eBay. "Some heinous collector robbed me of a Shakespeare Silent, a sweet green one," he said.

"Are you Gnarlmaker?"

"Doh! Was that you? You're driving up the price of these reels." He sounded a little peeved. Not long afterward, though, a small box showed up in the mail. It contained a gift from Sillett—a dented, banged up, automatic push-button fly-reel from the 1930s called the Miracle. He said that a Miracle with a grapnel hook "rules in the canopy." He had virtually cornered the market on Miracles. Over a period of a year or two, Sillett had bought twenty Miracles on eBay and at flea markets, and had stashed them away in his garage. In doing so, he had probably become the world's leading collector of Miracle fly-fishing reels, while steadily driving up the price of a Miracle from five dollars to, finally, twenty dollars, when he stopped buying them, because he thought they were getting too expensive.

SILLETT AND ANTOINE AND THEIR COLLABORATORS HAD ALWAYS HAD difficulty raising enough money to carry on their studies of redwoods. Like most scientists, they depended largely on research grants from government agencies and private sources. "In ten years, I've spent less than half a million dollars, in total, on research in the redwoods," Sillett said, not long after I met him. "It's a drop in the bucket compared to what other research programs can cost." The exploration of the redwood canopy was being done on a shoestring and with a certain amount of dumpster-diving on eBay. Sillett's graduate students climbed for about a hundred dollars a day, and the pros got nothing, mostly (they volunteered a lot of the time).

For some years Reese Halter, a philanthropist in Canada, funded a lot of the redwood-canopy research, but Halter eventually began to direct his attention to other areas. In 2002, Sillett applied to the National Science Foundation for a grant of $200,000. Steve's brother, Scott, had become an internationally known ornithologist at the Smithsonian Institution, and he had a lot of experience writing grant proposals. Scott advised Steve on how to write the N.S.F. proposal. Steve honed the proposal, hoping it would please a review committee in Washington, D.C.

"We have a decent shot at this," Sillett said to me on the phone one day.

The grant proposal was rejected. "I don't know what we're going to do," Sillett said. "At least Marie and I have jobs."

GEORGE KOCH, THE TREE PHYSIOLOGIST FROM NORTHERN ARIZONA University, is a lanky, genial man in his late forties, with knotted arms and an easygoing manner. Sillett taught him how to climb trees. "I'm like a kid in a candy shop, climbing these three-hundred-and-sixty-foot-tall trees with Steve," Koch said to me one day in Arcata. "The overwhelming question for me is what determines the height of a tree. At around three hundred and seventy feet, the tallest redwoods seem to be approaching a ceiling. The tall-tree ceiling is based on a limit to the height to which any plant can lift water. Why aren't the redwoods six hundred feet tall?"

Koch and Sillett wrote a paper about this that was published as a letter to *Nature,* the British journal of science, in April 2004. They had found what they believed are natural limits to the height of any plant, including the redwoods. They think that the height limit is somewhere around 420 feet. It seems that no plant can grow taller than that. The *Nature* paper got them a lot of attention.

In the aftermath of the *Nature* paper, George Koch and Steve Sillett applied to the National Science Foundation for a grant. This time they said they wanted to study the world's three tallest species of trees, to learn how they grow so tall, what natural forces act on plants to cause them to become gigantic, and to gather clues about what is happening to the gigantic plants as the earth's climate changes. They proposed studying the *Eucalyptus regnans* of Australia, the redwood, and the Douglas-fir. This would be the Tallest Trees Project.

To their surprise, the National Science Foundation awarded them $780,000.

When I heard about the grant, I called Sillett and Antoine to congratulate them and found that they were already starting to pack their gear for a climbing expedition to Australia. They were headed for the same *Eucalyptus regnans* groves where they had spent part of their honeymoon, on the Hume Plateau, in the Great Dividing Range. This time they would explore the canopy more thoroughly. They would select and climb fifteen trees. They would install permanent climbing systems in them, and they would measure the heights of the trees they'd climbed to the nearest inch. The trees would be studied for five years.

Sillett and Antoine had hired Jim Spickler, Sillett's former graduate student, to climb with them. They were looking for a fourth professional climber. But since Sillett had kept a pretty tight lid on his tall-tree climbing techniques, he had run into a shortage of qualified expedition climbers. All the other tall-tree climbers were busy with projects and weren't available.

"What about me?" I said.

SKELETON FOREST

THE MOUNTAIN ASH, THE *Eucalyptus regnans*, MAY ONCE have been the tallest tree in nature—taller than the redwood. It occurs in rainy mountains and valleys in the state of Victoria (in southern mainland Australia) and on the island of Tasmania. The mountain-ash tree is an angiosperm, a flowering plant. It is the tallest flowering plant. The tree has smooth, bone-colored bark, which turns pale green in the spring. In the fall, the bark turns brown and peels off the tree, and it hangs down along the trunk in strips that can be as much as a hundred feet long; the strips can persist for years, swaying and clattering when the wind blows. A mountain ash has droopy, pointed leaves. The leaves have a polished surface that reflects sunlight, giving the crown of a *Eucalyptus regnans* a glittery appearance. The main trunk is a straight, round bole, like ivory, with a smooth, polished surface. The trunk can be fifteen feet in diameter near the ground, as wide as a redwood, and it rises to incredible heights out of a fluted, star-shaped base. The trunk runs straight upward for as much as twenty stories before the first branches appear and the crown begins. The crown is an explosion of snaky limbs, and it typically holds a mishmash of dead hangers—limbs dangling or torn off and balanced on other branches. The crown of the tree looks like a skeleton with broken bones hanging in it, and is dangerous for climbers.

It is not clear which of the tree species that Sillett and his collaborators were studying (the redwood, the Douglas-fir, or the *Eucalyptus regnans*) is naturally the world's tallest living thing. In the late nineteenth century, some Douglas-firs were cut down that were reportedly more than four hundred feet tall: the Douglas-fir may once have been taller than the tallest redwoods. But, then again, there may have been redwoods back then that were taller than any Douglas-fir. In the late nineteenth century, there were reports from Australian loggers that some of the mountain-ash trees they felled were more than four hundred feet tall.

Nearly all of the virgin old-growth mountain-ash forests have been logged away. Most of the last remaining virgin stands are in Tasmania, but the trees there are being chipped up and turned to pulp. The pulp from a ground-up mountain ash has a fine texture and a clean, white color, and is used for products like toilet paper.

The tallest living mountain ash is named Icarus, and it grows in the Styx Valley of Tasmania. The tallest trees in the Styx Valley are largely dead in their tops. They are dying naturally, of old age, and so the tallest mountain-ash trees in the Styx Valley don't form an unbroken, green, living forest canopy. The tallest unbroken forest canopy in the Southern Hemisphere is a stand of *Eucalyptus regnans* trees at the center of the Hume Plateau, near Mt. Disappointment, at the southern end of the Great Dividing Range, about forty miles north of Melbourne. The trees on the Hume Plateau reach heights of between 290 and 300 feet. The climate on the plateau is cool, rainy, and wet, though it's not quite wet enough to be called a true rain forest. The Hume Forest has dry spells and bursts of hot, windy weather, when forest fires can break out.

The area where the tallest mountain-ash trees live is closed by locked gates and barbed-wire fences. It has been off-limits to visitors for more than a hundred years, and trespassers collect a serious fine if they're caught there. The tract is called the Wallaby Creek Designated Water Supply Catchment Area, and it is a source of drinking water for the city of Melbourne—the equivalent of a reservoir, except that it's a virgin forest. The plateau is two thousand feet above sea level, and is stuffed with vegetation and drained by impenetrable, traplike gullies. It is threaded by a handful of dirt roads marked "position doubtful" on government maps. The only people who are allowed into the area are park rangers and employees of the government-owned Melbourne Water Company, and a handful of biologists. Nobody goes there much, anyway. The forest is full of leeches.

STEVE SILLETT, MARIE ANTOINE, AND I DROVE IN A RENTED CAR ALONG a dirt road that climbed the flank of the Hume Plateau. It was a cool, blustery morning in January 2005, at the height of the Australian summer. Our car was packed with climbing gear. We had spent the previous night in a tin-roofed bungalow provided to us by a ranger at Kinglake National Park. We slept on the floor—the bungalow didn't have any beds. It also didn't have heat. The inside of the bungalow was forty degrees Fahrenheit, and we wore wool hats and warm clothing indoors. I was surprised by how cold summer in the mountains of southern Australia can be. As we followed the road upward onto the plateau, raindrops began splattering on the windshield.

"The darkest part of the sky would be sitting over the plateau, of course," Antoine said, peering ahead.

"It's going to be as slick as snot up there," Sillett said.

The road went through a forest of eucalyptus called stringybark messmate trees. A wallaby loped out of some ferns, stopped, stared at us, and hopped away. At an altitude of exactly eighteen hundred feet, the messmate forest switched to *Eucalyptus regnans*: suddenly we were in the mountain-ash groves.

At the top of the plateau, the mountain-ash forest was a house of light, wind, and sound: utterly different from a redwood forest. The trees were pale, ghostly columns soaring twenty-five to thirty stories overhead, flaring into craggy, distant crowns splashed with sunlight. When the wind blew, the bark that hangs along the trunks would clash and clatter, making eerie noises. Sulfur-crested cockatoos flew from tree to tree, screeching, and occasionally a flash of brilliant red would soar on a diagonal among the trunks, a parrot called the crimson rosella. The trunks ran up from a dense green, junglelike under-story, a layer of acacia trees, musk daisies (treelike shrubs), Austral mulberry, flowering Pomaderris trees, and tree ferns. The tree fern is an ancient kind of fern that looks slightly like a palm tree. It was widespread in the age of dinosaurs. The tree ferns on the Hume Plateau can grow to more than thirty feet tall. The musk daisy tree-shrubs are related to the daisies and asters that grow in meadows in the Northern Hemisphere. In Australia, the daisies have evolved into small trees.

The understory of the Hume forest forms a thick lower canopy that's fifty feet above the ground—a canopy that is the same height as many stretches of woods in the eastern United States, Europe, and Japan. This lower canopy is just the beginning of the Australian forest. The mountain-ash trees rise through the lower canopy in spectacular columns, and mushroom into an upper canopy far above the first, like a layer of high clouds. We parked near the center of the plateau, put on our backpacks, and bushwhacked into the tree ferns. We were heading for a tree named Big Ash Two, which is the second-tallest tree in main-land Australia. It is also named Amabilis, which is Latin for "lovely."

THE DISCOVERER OF AMABILIS, AND OF VIRTUALLY ALL THE TALLEST trees in the Southern Hemisphere (they are mostly *Eucalyptus reg-nans*) is a soft-spoken classical guitarist and high-school music teacher named Brett Mifsud. Mifsud, who teaches in the Croydon school district, east of Melbourne, is a lean, quiet man in his late thir-ties, with curly brown hair and a craggy face. Steve Sillett describes Brett Mifsud as the Michael Taylor of the Southern Hemisphere. Mif-sud has long fingernails on his right hand, for plucking the strings of a guitar. In the early 1990s, he was living in a cabin in the mountains and earning a living by playing classical guitar in bars. There was not a huge demand for this, and he had some time on his hands and no money. He got an old bicycle with no brakes and began exploring the Wallaby Creek Catchment and following the uncertain roads. He didn't think the maps of the area were accurate or complete, and he was looking for waterfalls that weren't noted on the maps; he found several lost waterfalls. "The forest at Wallaby Creek guards its secrets," Mifsud said to me. As he bushwhacked, he picked leeches off himself and burst them with his sharp fingernails, and he tried to keep them out of his eyes. "You don't want to get a leech in your eye," he said. "It makes you look like Alice Cooper."

He loved the trees. He started carrying a clinometer with him. This is the same device that Michael Taylor built for forty-five cents when he first started measuring trees. Studying the mountain-ash trees and estimating their heights, Mifsud soon realized that he was discovering unknown, unnamed, world-record giants, but their heights remained uncertain because his instruments were crude. He also went to Tasmania, and he explored the Styx Valley, where he discovered many of the tallest mountain-ash trees known there. He began finding large mountain ash—not the tallest but the biggest. He named one of them the Hairy Godmother.

One day in 1996, Brett Mifsud was surfing the Internet and he noticed a website run by Michael Taylor, who described his discoveries among the redwoods of California. Mifsud wrote a long, emotional, handwritten letter to Taylor, telling him about the giant trees he was finding in Australia. They soon became friends. In 2000, with earnings from his job as the Mr. Fixit of the Nicoderm-patch machines, Taylor bought a top-quality laser range finder and mailed it to Mifsud as a gift. ("Brett was hurting for money at the time. I wasn't," Taylor said to me.) Mifsud began using the laser to obtain accurate measurements of the trees on the Hume Plateau. Brett Mifsud got to know a Melbourne arborist named Tom Greenwood, who is one of Australia's top tree climbers. Mifsud showed Greenwood the tallest trees to climb. Greenwood got a measuring tape and began measuring the trees as he climbed them, and the two men combined their data. Brett Mifsud's high-school guitar students had no idea that he was getting an international reputation in tall-tree botany.

WE PILED OUR GEAR NEAR THE BASE OF AMABILIS. THE TREE HAD been climbed only once, a year earlier, by Tom Greenwood. Amabilis contained tons of hanging dead wood. We had no clear information on what Greenwood had found in the tree when he entered its crown.

HOUSE OF LIGHT. The Hume Plateau. *Drawing by Andrew Joslin.*

Sillett got out his bow. The wind grew stronger. There was a clacking sound—the strips of bark dangling from the trees were banging around. The southern end of the Great Dividing Range, where the Hume Plateau is situated, stands in the path of winds coming off the Bass Strait and the Southern Ocean, between Tasmania and mainland Australia. That stretch of ocean has some of the wildest seas anywhere in the world. When the wind roars off the Bass Strait, the first thing it hits is the Hume Plateau. One day Sillett asked Greenwood whether he should be climbing *Eucalyptus regnans* in a strong wind. Greenwood told him that it was dangerous even to be underneath the trees. "When the wind blows, get the hell out of that forest," he advised Sillett.

Sillett leaned backward, staring upward. "Look at the trees dancing up there. We're going to get a ride today," he said. He drew and fired an arrow. There was a pause of seconds while it lofted higher and higher—fifteen stories, twenty stories. The wind caught the arrow and it went astray, falling down along the trunk. "God, these trees are intimidating," he said.

Antoine was standing off to one side, keeping her eyes on him. She had an inscrutable look.

The rain stopped, and the air temperature soared, going from about fifty degrees Fahrenheit to eighty within the space of a few minutes. We began to sweat, and the tree ferns began to steam. I took off my rain jacket and placed it on the ground, and sat on it. But not for long. All of a sudden, the jacket had hair on it. Dozens of leeches were traveling across it, moving like inchworms. The species was the Australian land leech. It does not live in water, as most leeches do—it lives on the ground and on twigs and leaves.

I got to my feet, holding the jacket up by two fingers, and started picking leeches off it. The leeches, I discovered, were sticky. I tried to flick them off, but they stuck to my fingertips. I tried to shake them off, and rub them off, but they crawled down my fingers, down my wrists, down my forearms, heading inside my clothing. I tried to crush them with my fingernails, but they were leathery and as tough as gristle. A dozen leeches began moving along my arms. Then I noticed that my pants had developed mobile hair. They were climbing up my trousers, heading for my crotch. I put tick repellent on my pants, and that seemed to confuse them, but not for long. When they encountered the repellent, they hesitated for a moment and then crawled over it. Abruptly I felt them on my ears and neck. These leeches had fallen from above. They were hanging by the hundreds in the ferns and leaves. The instant they smelled us, or sensed our body heat, they would let go and fall on us.

"Interesting place," Sillett said, picking a leech off his thigh. He leaned with his back against a tree, to steady himself, nocked an arrow, drew his bow, and fired, squinting. It flew into Amabilis, with a fishing line streaming off a reel on the bow. The arrow vanished. He grabbed a pair of binoculars and tried to see where it had gone. There was a leech on his forehead. The arrow had gone through the crown, but the fishing line hadn't passed over any decent branch; it was hanging over twigs.

Sillett swore. "One trick for getting an arrow correctly through a tree is to be extremely stubborn about it," he said. He reeled in the line, aimed again, and fired, and we watched the arrow go higher and higher. This time it went true, and sailed over a strong-looking branch in the center of the crown, about twenty stories up. "Golden shot," he said.

"Yeah, and I'll bet it went over a hollow branch," Antoine said dryly.

"Everything's hollow up there."

Antoine took off her helmet and began de-leeching the inside of it.

Sillett stared at his bow. A leech was crawling along the bow-string. "There's something so sinister about the concept of a leech," he said.

We got a rope pulled up over the branch, and I tied one end of it around the base of a nearby tree. The rope ran up two hundred feet, was draped over the branch, and hung loosely down to the ground. I double-checked the knot that I had tied on the ground anchor, since Sillett's life would depend on my knot. Sillett would go aloft by climbing up the hanging end of the rope. Antoine and I would follow him as second climbers. Antoine faced him and inspected him—a safety check.

"The wind's picking up," he said to her in a low voice. He bent over and tightened something in his harness. "But it's not too bad." A smile flickered, and he looked around. "I wonder what I did with the car keys." He always lets people know where the car keys are before he begins an ascent of a potentially dangerous tree. There was a roaring sound, and the crowns of the ash trees began pulsing. "It's picking up again. It might be suicide up there."

Antoine picked a leech off his collar. "If you think it's suicide, Steve, there's no way I want you climbing," she said.

"Well, I'm going to climb."

"Okay, but I want you to be careful."

"There's nothing to worry about, unless a branch breaks."

"A big 'unless.' "

They kissed. Antoine seemed subdued.

Sillett bent over and inspected his pants. "I think I'm leech-free." He clamped his ascenders to the rope and began jugging into Amabilis. He was hanging in the air, far away from the trunk, spinning in the wind as he ascended.

The wind slacked off, and the tree grew quiet. Sillett turned into a struggling homunculus, moving into a skeleton forest, outlined against the sky. There was a sharp cry: *Wheeeooowhip!* The whip bird. Then there was a creaking sound, like a door opening on dry hinges. It was the call of the gang-gang, a type of parrot. I contemplated my fiberpile jacket. It was flecked with squirming boogers: the leeches could crawl across dry fiberpile—unbelievable. A sour smell drifted in the air. It came from a stink-ant nest somewhere nearby.

Marie Antoine had walked back away from the tree, to a spot where she could see her husband clearly as he climbed. She held her radio in her hand, ready to speak to him at any moment. "It's the initial climb upward that scares me," she said. "I worry that something could fall on him."

Antoine never takes her eyes off Sillett when he's ascending a tree, ever, as long as she can keep him in sight. She stands with her hands straight at her sides and her face turned upward, with a searching look.

■ ■ ■

THREE MONTHS EARLIER, SILLETT HAD BEEN MAKING AN ASCENT INTO a wild, unexplored redwood in southern Humboldt County, and something went very wrong. As we watched him climb Amabilis, she told me what happened. "I didn't like the tree to begin with," she said. "Some trees are disturbing, and this one gave me a bad feeling. When he started to climb, I could see a big, dark silhouette of something flopping around at the top." She ordered him to abort the climb—"Get down *now.*"

He obeyed her, and rappelled to the ground, fast. Just as his feet touched the ground, the object came loose.

"There was a sickening sound," she recalled. A five-hundred-pound hanger, as heavy as a Steinway piano, appeared, falling from near the top of the tree. "Headache!" she screamed.

As the widow-maker fell, it got caught on Sillett's climbing rope, and the rope guided it down toward him. He saw it coming. He unclipped himself from the rope and ran for his life. The branch fell twenty-five stories, twirling around the rope, and hit the spot where Sillett had been standing a moment earlier, with an impact that made the ground shake. Afterward, they threw the rope away—they didn't know what kinds of stresses and strains it had been under when the hanger fell along it, and it could never be trusted to hold a human life again. Marie Antoine wouldn't forget the incident anytime soon.

The sun came out; the clouds vanished. Hot white light beat down on us, and a crazy laugh echoed around, a wacky-sounding chortle. A kookaburra bird.

"*Kookaburra sits in the old gum tree, eating all the . . . ?*" Antoine sang, and she stopped. "Hm. What does he eat? Gumdrops, I guess."

"Are you and Steve planning on having children?" I asked.

Still staring into Amabilis, she said, "I'll be turning thirty next year. At that point, I'll decide."

"Marie, I'm off the main rope," Sillett said on the radio. "You can go ahead and untie the anchor."

"I will let you know, Steve, when that is done."

Their communication by radio was precise. A misunderstood word could result in someone's death.

She went over to the place where I had tied the rope. "The anchor is untied, Steve," she radioed.

Up in Amabilis, Sillett clipped the climbing rope to himself and went into free motion using his spider rope. As he climbed, he would drag the main rope along with him and re-anchor it near the top so that Marie Antoine and I could climb up it. The wind kept rising and falling. The crown bent, folded into itself, and expanded.

"He's swinging in the bluster," Antoine said, holding her hand to shade her eyes.

In less than half an hour, he had spidered his way to the top. He tied the middle of the main climbing rope to the tree and let both ends hang down along opposite sides of the trunk.

Amabilis was surrounded by a debris ring—a tangle of broken and rotting limbs, lying on the ground. They had accumulated in recent years, falling out of Amabilis's crown. Antoine and I clambered up on the debris ring, picking our way across greasy, slick logs. I clamped my ascenders to one rope, and she clamped hers to the other, and we went aloft, jugging steadily, occasionally spinning on our ropes. The wind picked up, and a sheet of zinc moved below the sun and the tree began to sway. The wind ramped up into a serious roar. "It's getting that freight-train sound," Antoine said. The tree was swaying. Sillett was far above us, nowhere to be seen.

We passed a huge, rotting limb, snapped off and hanging by shreds. It was lying vertically against the trunk, and it looked as if it could let go and fall at any moment. I felt as if we were picking our way around the business end of a glacier. "In these *regnans* trees, it's the large live branches that are the ones you have to worry about, because they're actually the ones that break off and fall, maybe as much as the dead ones," Antoine said. In other words, the whole of Amabilis was a hanging glacier.

We stretched a measuring tape around the trunk every ten meters as we climbed, handing the tape back and forth between us, and maneuvering in the air on the ropes. Antoine read off the numbers to me, and I wrote them down.

When canopy scientists want to travel in a circle around the trunk of a large tree, they swing like a pendulum. Holding the end of the measuring tape, I planted my feet against the tree and pushed off. I drifted a considerable distance outward, floating away from the trunk, and as I went out into the wind it tugged at my clothing. I looked down. On the forest floor below, the tops of tree ferns looked like green stars, and I could see a tiny backpack. The wind caught me, and the ferns began turning around—I was turning in the wind. I drifted back to the tree and kicked off again, harder, and drifted farther away from the tree, while the measuring tape fluttered in my hands. By pushing away, and swinging, I bounce-walked in a circle around the trunk, grabbing pieces of bark with my fingers. It was, perhaps, like walking on an asteroid, where there is only slight gravity.

The tactical rope on which I was hanging was three-eighths of an inch thick, the size of a person's pinkie. It ran straight up into nothingness and vanished, a tense dark thread. It sang in the wind, and stretched and bounced slightly, like a long rubber band. Antoine was carrying a second measuring tape, a large reel of tape. She had attached one end of this tape to the ground, and she was unreeling it as she climbed, so that we could measure the height of Amabilis to the nearest inch.

The tree had epicormic branches popping out of its trunk. Bounce-walking around, I used a positioning lanyard (the short monkey-tail-like rope) to hook myself to the epi branches and pull myself around the tree. I slid the measuring tape underneath strips of bark to get a more accurate measurement of the tree's diameter.

"You want to be alert when you put your hands underneath the bark," Antoine remarked. "It's where the big spiders like to hang out."

It was an unidentified species of canopy spider that she had noticed during her honeymoon in the Skeleton Forest. It was possibly the huntsman spider, though no one was sure about that, because no one had dared to try to collect one. The spider seemed as big as a tarantula. It had fangs, and it lived in holes and under shreds of bark. She had seen these spiders moving very quickly. ("They're freaky fast, and they've got teeth," as Sillett put it.) At least there are no leeches in the Australian canopy.

I lowered myself down the rope a few feet and peered under some hanging bark.

"If you shine a light in there, you might see eight little purple eyes shining back at you," Antoine said.

I didn't see any eyes. I pushed the measuring tape under the bark and pulled it through, and kept swinging around the trunk.

In 1949, David H. Ashton, a young Australian biologist, began studying the mountain-ash forest in the Wallaby Creek Catchment. Ashton would spend more than fifty years studying the forest of the Hume Plateau. He discovered that all of the *Eucalyptus regnans* trees at the heart of the forest are precisely the same age. They were all born within about two weeks of one another, in the summer or fall of 1710. (Ashton learned this by counting the growth rings in some of the trunks, and comparing the rings very carefully.) The forest burned sixty years before Captain James Cook claimed Australia for Great Britain. The seeds of *Eucalyptus regnans* need a forest fire in order to germinate. Fire shapes the forests of Australia. In 1710, the heart of the Hume Plateau burned to a black wasteland in a great fire. The generation of trees on the plateau sprang up immediately after the fire, from charred ground, and from seeds that had been dropped by their parents before the parents burned to death.

The canopy we had entered was close to three hundred years old, and it was beginning to die. Because fires frequently burn mountain-ash forests, the trees don't naturally live very long. They don't invest many resources into fighting off disease and insects, because a fire will probably get them first, anyway. They've evolved a strategy of growing fast and very tall, producing lots of seeds, and dying young. The *Eucalyptus regnans* is one of the fastest-growing living things known. A young mountain ash can grow six to nine feet taller in a year, and it can grow to be two hundred feet tall in seventy years. When they reach a certain maturity, mountain-ash trees slow down in their growth rates and seem to get tired. Their limbs begin to rot and die and fall off, the branches become hollow, the tree's crown develops dead areas, and often the entire trunk of the tree rots from the inside. Mountain-ash trees rarely live past about four hundred years. When a mountain ash dies, it often dies spectacularly and suddenly and in a cataclysm of violence, like a redwood, but the manner is different. A *Eucalyptus regnans* often dies from what is called a crown failure.

In a crown failure, the tree's crown suddenly implodes and collapses inward and down around the trunk. This leaves a broken cylinder, which is the trunk, sticking up as much as a hundred feet into the air, dead. It can look like a broken obelisk. It is surrounded by a debris ring, the remains of the crown. The Hume Forest is full of obelisks standing in debris rings. They are the casualties, thus far, of the generation of 1710, and most of them have died in the winds that punish the Hume Plateau.

WE HAD GOTTEN ABOUT HALFWAY UP AMABILIS WHEN THERE WAS A cracking noise overhead. We heard a big shout, and it sounded involuntary. The top of the tree shook, and our ropes trembled. We heard cursing, and the *F* word was launched into the canopy.

"He's making scared-Steve sounds," Antoine said. She got out her radio. "What is it, Steve?"

"Nothing." Pause. "Headache," he said. A branch fell past us, swishing as it fell.

I got on my radio. "Steve, what happened?"

"I broke off a branch. I was anchored to it, and it peeled off."

"How far did you fall?"

He had taken an eight-foot whipper out of the top of the tree. He had been hanging on a branch from his spider rope and holding a laser range finder in his hands, trying to measure the top of the tree, when the branch suddenly snapped off, and he plunged through the crown. The other end of his spider rope had been wrapped around a solid branch, lower down. He had fallen until he was jerked short by this second anchor, and he ended up swinging from it, spinning around and cursing.

When we got to the top, we saw that many of the branches had holes in them and were hollow. Bright-red sap oozed from some of the holes—the sap of the *Eucalyptus regnans* is dark crimson in color. "There's nothing like climbing a hollow shell that's ninety meters tall and filled with blood," Antoine remarked. She handed the measuring tape up to her husband, who was above her. He turned himself upside down on his rope, hanging like a bat, and grabbed it and righted himself. Amabilis was 90.85 meters tall—298 feet and three-quarters of an inch.

Insects were crawling on the branches and on the leaves. I plucked one off and looked at it. It was a species of wingless leafhopper, unable to fly, unable even to hop, apparently. It crawled tenaciously around on my hand. I could feel its legs picking at my skin. I tried to pull it off and put it on a leaf, where I was sure it would have a better life than on me, but it wouldn't let go of me. The leafhopper was a denizen of the high Australian canopy, very probably an undescribed species. (Nobody has looked at the insects of the old-growth mountain-ash canopy.) The leafhopper, being flightless, and living very high above the ground, had apparently evolved a behavior of clinging fiercely to surfaces and never letting go. It was pretty clear that if the leafhopper ever fell from the tree it would be a goner on the forest floor, twenty-eight stories below. I finally managed to place it on a twig.

I threw my spider rope over a branch and clipped myself in, and swung into the air a short distance, and looked out. Cloud streets were moving: wind squalls rolling in from the Bass Strait. We could just make out the towers of Melbourne in the distance. The mountain-ash trunks were long gray shafts that sprang out of the lower canopy, extended up, and arched out into tangles, like the columns in the nave of a cathedral that rise out of their pediments and join the voissoirs of the cathedral's ceiling.

The ash canopy extended in all directions across the Hume Plateau, opaque, tangled, unknown. I tried to add up the names of the people I knew who had climbed into this canopy so far—they included Michael Taylor. The forest canopy of the Hume Plateau had been explored by just ten people, including ourselves.

Back down in leechland, we pulled the rope out of Amabilis, having made some precise measurements of the tree and installed a climbing system in it.

As we were organizing our gear, Sillett stepped out of his harness and glanced down. His crotch had a dark stain. It was blood. "Uh-oh," he said, and he turned his back and opened his pants and looked inside. "Oh, God," he said. Several leeches had climbed Amabilis with him.

"What's the matter, Steve?" Antoine asked.

"I think I got my period."

Jim Spickler arrived the next day, and we formed ourselves into two climbing teams and climbed the fifteen mountain-ash trees, measured them and rigged them with climbing systems, and went home. I continued to assist with the Tallest Trees Project once in a while in Northern California, working as a professional though unpaid climber.

JOURNEY TO KRONOS

WHEN I FIRST MET STEVE SILLETT, HE MENTIONED THAT the top of Kronos, a tree that stands at the west end of the Atlas Grove, had broken off and was resting against the top of Rhea, beside Zeus, the tallest member of the Atlas Grove. In Greek mythology, the Titan Kronos was the father of Zeus and the god of time. Eventually, the broken top of Kronos had slipped loose from Rhea and calved to the forest floor. Broken limbs and pieces of shattered buttresses had rained down through both trees, leaving widow-makers and hangers suspended throughout their crowns. The event may also have left debris hanging in Zeus—no one knew, because nobody had dared to climb up and try to find out. The calving event left a tangle of rigging lines jammed in the trees' crowns, mixed with pieces of defunct scientific gear.

Two years after the top of Kronos fell, a strong winter storm hit Prairie Creek Redwoods State Park and knocked over several large redwoods. The storm did further damage to Kronos and Rhea.

The west end of the Atlas Grove had been abandoned by researchers for years. In the winter of 2006, Sillett and Antoine decided to enter the area, explore it, and remove the damaged rigging and scientific gear as well as all other traces of human presence. "I don't like to see techno-trash stuck up in redwoods," Sillett said. They asked me to come with them. We would try to restore the trees to a wild state, and after that they would not be climbed again.

Since Rhea and Kronos were full of concealed debris, the best way to get into them, it seemed, was to climb up Zeus and then attempt to skywalk across the top of the canopy into them. "We're going to go over to Rhea and Kronos from the top and climb down through them, so we won't be exposed to something falling on us from above," Sillett explained to me.

Movement from tree to tree is called an aerial traverse. A climb that proceeds in multiple aerial traverses horizontally through a canopy is called a canopy trek. In canopy trekking, you pull your ropes and gear along with you as you skywalk through a forest, leaving nothing behind. We would carry very little rope, and the least amount of gear possible, so that we could move fast and lightly.

ON THE MORNING OF OUR PLANNED CANOPY TREK, RAINS WERE PREdicted with high winds; the weather radar showed that a storm front was moving in from the Pacific Ocean.

"I'm not a wuss," Marie Antoine said. "Well, I'm somewhat of a wuss. But I'm ready to climb today."

We bushwhacked into the Atlas Grove carrying our gear. It began to drizzle, and the sky got very dark. The ground at the west end of the grove was choked with freshly fallen limbs, buttresses, and branches, mixed in with fresh green foliage. Much of the debris had fallen from Rhea and Kronos within the past few weeks, and some might also have come from Zeus—we couldn't tell.

"We'll just have to see what the situation is when we get up there," Sillett said, peering up into the spaces among the trees.

We pulled a rope into Zeus. Nobody had climbed Zeus since the day Sillett and Antoine had gotten engaged there.

As Sillett climbed up the rope, Antoine stood on a log and

watched him, holding a radio. The rain began to fall more steadily. The sea was rumbling and groaning beyond a ridge, five miles away. A loud, raw winter sea.

Sillett climbed slowly, with extreme caution. The fresh chunks of redwood that lay on the ground amounted to the largest debris ring I had ever seen. There was no question that the sound of this stuff falling would have carried for a mile or two.

Sillett got near the top of Zeus, and pretty soon two climbing ropes were hanging down, and Antoine and I went aloft.

NEAR THE TOP OF ZEUS THERE IS A BONSAI BUCKTHORN, AND NEAR IT we held a climbing conference and made an inventory of our ropes. We were carrying a total of four spider ropes. They had different colors and were of varying lengths. Antoine had brought a standard sixty-footer with yellow and green stripes. Sillett had two spider ropes: one was a long dark-green rope, good for throwing long distances, and the other was blue and much shorter. My spider rope was white and the shortest of all.

I had some misgivings about my rope. It was only fifty-five feet long, a little too short for use in redwoods. Because a spider rope is doubled into a loop when it is passed over an anchor point, the maximum reach of the rig is only half the total length of the rope. The maximum distance that I could move through the air from one anchor point to another was only twenty-seven feet. If I encountered any empty zones in the canopy where the anchor points were more than twenty-seven feet apart, I would be in a jam and we would have to abort the trek and try to go back, which wouldn't be easy. Certainly my spider rope wasn't long enough to enable me to skywalk from tree to tree. In fact, as we considered it, none of our spider ropes were long enough to allow us to get to the ground from the lowest branches of Rhea and Kronos. The lowest branches on those trees were roughly 150 feet above the ground, and not even Sillett's long green rope would extend that far. We could move across the canopy, but how would we get down?

"We'll figure something out," Sillett said quietly to Antoine.

There was a rushing sound, and Zeus began to move ever so

slightly, like slow breathing. "Here comes the wind," Antoine said, looking around.

Sillett and Antoine thought that with a rising wind and a storm predicted, and given the small amount of rope we were carrying, it would be wise to have one member of the team serve as a backup climber. I didn't know enough about canopy trekking to know what a backup climber does. They explained that the backup climber stays within easy reach of the ground, and is available to help the other climbers make an escape from the canopy, if necessary.

I had a feeling that I would like to be stationed close to the exit. Antoine, however, volunteered to be the backup climber. "I'm content to chill," she said. She would stay behind in Zeus, while Sillett and I crossed into Rhea and Kronos. As we explored those trees, she would shadow us in Zeus, staying at our height and keeping us in sight if she could. If we got into trouble, she could help us get a rope back into Zeus, so that we could escape by making a lateral traverse into Zeus and then down, or she could go in search of help, if that became necessary.

SILLETT WAS STANDING NEAR THE TOP OF ZEUS, FACING RHEA. He walked out on a branch as far as he could, unfurled his long green rope, and flung it toward Rhea. He got the rope draped over a branch in Rhea, dangling down. Then he pulled a grapnel hook from the fly-fishing reel on his saddle and tossed the hook toward the end of the green rope. He snagged the rope and pulled it back to him, then clipped himself to the loop, stepped out into space, and crossed over to Rhea, hanging from a V formed by the blue and green ropes, one anchored in each tree.

Sillett disappeared inside the top of Rhea. We could hear him moving around. "There's a giant hanger here," he called out. He reappeared, walking along a slender branch nearly out to its tip, dragging a black mass. It was a piece of Kronos. He wrestled it out of the tree and kicked it into space. It landed in the debris ring with a crash.

It was my turn to cross over. Since my spider rope wasn't long enough to carry me from Zeus to Rhea, Sillett proposed using his green rope to construct a zipline, a horizontal traverse rope. In order to set up the zipline, I would need to get a line over to him. I was carrying a bag that contained a long yellow throw line with a weighted throw bag on the end of it. I unfurled the line and tossed the bag toward Sillett. It traveled on a high arc, like a pop fly, dragging the line after it, and landed in Sillett's arms.

We used the throw line to establish the traverse rope, and I clipped a pulley to it and bobbed toward Rhea. Halfway across, I stopped and looked down. I was thirty-one stories up, but there was no sense of height: I was hanging a few feet above floorlike masses of greenery.

I came into Rhea and anchored myself to a branch. Sillett untied the zipline and pulled it out of Zeus. It was like casting off a line from a ship. The rope had been our last connection to Zeus and to the main rope leading to the ground. We had gone into motion.

Over in Zeus, the backup climber stretched herself out on a branch. The weather seemed to be improving.

Sillett and I spidered a short way down through the top of Rhea and came into a small, beautiful glade. Rhea had a double top, with twin trunks, and there was a garden between them. Rhea Garden was a deep pocket of soil that had become established near the top of the tree. The pocket was filled with plants, mosses, and lichens. It was like a tiny Japanese garden, and it was probably as old as the Muromachi tea gardens of Kyoto. It was built up of layers of earth that had drifted into the tree. I did not want to touch it.

A curving redwood branch arched over the garden like a bridge. I branchwalked along the bridge, keeping my weight suspended on my rope as I looked down into the garden. It was enveloped in shade and hidden in the surrounding foliage—a green chamber in the air.

Sillett had disappeared through a hole in the wall of the chamber. "Are you coming?" his voice filtered up through the hole.

I spidered down through the hole, went around a corner, and came out into a canopy canyon, the most spectacular view I had ever seen in the redwood forest. The canopy canyon was a blasted, ruined void between Kronos and Rhea, extending down into darkness. This was where the leaning top of Kronos had fallen. All the buttresses, hanging fern gardens, and huckleberry thickets had been torn off along both trees. What was left was a cleft filled with sunlight. Kronos Chasm. Debris and broken branches were suspended and clinging to both trunks all the way down into the chasm, as far as the eye could see, and there was no sight of the ground.

Sillett was standing sideways in the air with his feet planted on Rhea's trunk, suspended from a spider rope. He took up his radio and spoke to Antoine, who was somewhere out of sight in Zeus. "Everything is changed up here, Marie."

"I'm not surprised to hear that."

"It's virtually unrecognizable."

I sat down on a branch and tied myself to it using a very short rope. I was feeling dizzy. Below me was nothing but the ruined face of Rhea descending into darkness. In front of me, at eye level, was the sheared-off top of Kronos. It was a broken shell of rotting wood, with an interior that looked like sawdust. A single live branch stuck out of the shell.

Sillett tossed his rope across to Kronos, aiming for the branch at the top. He caught it with his rope, used his grapnel hook to retrieve it, and skywalked over to Kronos. He ended up standing on the branch.

"I can see the whole top of Kronos swaying under your weight," I said.

"What's it doing, vibrating?"

"Yes. I think you should be very careful."

"I'm not worried," he said, adding that if Kronos's top collapsed he could bail out—he could jump and swing back to Rhea on the green rope. "Do you want to come over here?"

"I think I'll stick to Rhea," I said.

He nodded and swung himself down below the branch so that he was hanging from it on his blue spider rope. He peered into a hole in the top of Kronos and took out his radio. "Oh, my goodness," he said. "This top is totally hollow. I can see all the way down through the middle of the trunk. We've found a rotten shell here."

Antoine's voice came over the radio: "Steve, I don't want you taking chances. Kronos is truly crumbling."

The rotten top of Kronos was crisscrossed with threadlike roots. A coast redwood seems to have the ability to send out roots from any part of its tissue, including its top. Kronos may be putting roots into its own dead top, and feeding on its own decay—as if a person who was dying of gangrene were able to get nutrients from his rotting flesh, and thereby keep himself alive somewhat longer.

Sillett slipped his anchor and cast off from Rhea, then he lowered himself down through Kronos Chasm along the rotten spire of Kronos, poking into the holes and looking at things. He flagged me on the radio. "Why don't you descend Rhea and clean up as much scientific trash as you can?" he said. "When you get lower down, you can cross over to Kronos."

I was still sitting on the branch, looking down into Kronos Chasm. Where was the way down? There were no branches. Rhea was just a scarred cylinder, and it was about twenty feet around at my height.

I unclipped my safety rope from the branch and swung out over Kronos Chasm, dangling from one end of my spider rope. I lengthened the noose and began descending along the face, looking for something lower down that I could attach my spider rope to. As I went down, I threaded my way among broken branch stumps, swinging back and forth, looking for a solid branch. I came to a hanger, a broken-off limb partly attached to the tree, dangling vertically by some splinters. I kicked at it with my foot to see if it was loose. It wasn't, so I swung under it. I passed through a fuzz of epicormic branches. Below them, the forest floor came into view—tiny stars of sword ferns scattered in a debris zone that was crisscrossed with great fallen spars of redwood, like the masts of shipwrecks, far below. To my left, clockwise around the trunk, a deadly-looking structure came into view. It was a tangle of widow-makers that reminded me of a beaver dam, suspended in a basketlike arrangement of limbs growing out of Rhea. The beaver dam consisted of pieces of Kronos and Rhea.

I kicked off and pendulated the other way, trying to stay away from the hanging beaver dam. A fat-looking branch came into view below.

Antoine spoke. "Do you see that branch below you? Try to get to it." She was slightly above me in Zeus, looking across the space between the trees. She had not been chilling. She had been watching us.

I thanked her and ended up just able to reach the branch. I anchored myself to it, released my first anchor, and swung out into the air, descending farther along the tree and hunting for the next available branch. A number of metal tags had been attached to the bark with nails. As I moved along, I pulled the nails out and put the tags and nails in a bag at my waist.

I spidered around Rhea, traveling in a downward spiral, until I lost sight of Sillett and Antoine. I was using the spider rope as an extension of my arms, grabbing branches with the ends of the rope as I maneuvered, holding on to Rhea with ropes running out of my hands, with my body dangling in the air.

Many of the tree's limbs had one or two salamander shelters buried in soil on them, bits of wood and plastic screen. I stopped on branches and lifted the things out of the canopy soil, put the screen pieces in my bag, cast away the bits of wood, and gently tucked the canopy soil back in place, smoothing it over and restoring the soil to its original state. Orange pieces of plastic surveyor's tape had been tied to twigs and branches in various places, evidently to mark the locations of the shelters. I began branchwalking around, pulling off the pieces of tape. A strip of tape was stuck to a small widow-maker hanging near the edge of the outer crown, where I couldn't branchwalk to it. I lassoed the widow-maker with my throw line and jerked the line; the widow-maker broke in half, and the pieces fell to the ground along with the piece of tape.

Sillett's voice came on the radio. "There's a really big hanger above you and to your right," he said.

"Where?"

"Keep going around. You'll see it."

It was a snapped-off branch, twenty-five feet long, hanging down vertically along Rhea. It easily weighed a quarter of a ton. I would have to crawl underneath the bridge in order to proceed downward. I wondered if it was stable.

"You want to check out that hanger before you go below it," Sillett said.

I hooked my rope over a branch above the hanger and climbed up to have a closer look at it. It seemed solid, so I spidered down and got on my knees and crawled underneath it, where I found a fern garden. I stood up and tiptoed through the garden, trying not to touch the plants, and kept going down.

• • •

AT A HUNDRED AND SEVENTY FEET, I STOOD ON THE LOWEST BRANCH of Rhea. It was an epicormic branch—a dog's hair branch, ready to be shed by the tree at some point—but I thought that it felt solid. The trunk of Kronos stood about forty feet away. Sillett was hanging in midair across from me, suspended from his spider rope, which was anchored over a lump of bloated gnarl as big as a sofa, which bulged out of Kronos above him.

"Why don't you come over here?" he said.

My spider rope wasn't long enough to reach across the space.

He threw me the end of his green rope, and it landed in a spray of small branches ten feet below me. I got out my grapnel hook and fished up the rope, clipping it to my harness. Then I looped my spider rope over the epicormic branch, hung my weight on it, and stepped out into space away from Rhea, hanging on a V of ropes between the two trees as I skywalked across Kronos Chasm. I ended up dangling in my harness in the air next to Sillett.

"You realize that you placed your life on an epicormic branch," he said.

"It was solid. Anyway, you've been known to do it."

We rested, getting ready for the final move, which would be downward through Kronos. The rain and the wind had stopped. The storm had never come. "We got a gift," Sillett said.

I could see Marie Antoine across from us, hanging in the air along Zeus. "The light's going," she reminded us.

Sillett didn't seem ready to go down. He was silent for a while. I could tell that these trees had a personal meaning for him. We were hanging within sight of the tree where he and his wife had declared their love for each other and had planned their future together. We were surrounded by the trees of the Atlas Grove, the center of his career as a botanist and a place in which he had risked his life many times. "There's always a moment during a climb when you lose yourself," he said. "You don't have a name anymore. When you find yourself in a place in nature where if you make a mistake you will die, you become open to what's around you. You start feeling the limits of your perceptions as a human being. You perceive time more clearly in red-woods, and you see time's illusory qualities."

We were looking down into a vast bowl of ferns out of which the three titans grew. "When you feel one of these trees moving, you get a sense of it as an individual," Sillett said.

"Do you really think of a tree as a kind of entity?" I asked.

"It's a being. It's a 'person,' from a plant's point of view. A tree is not conscious, the way we are, but it has a perfect memory. This is be-cause the trunk of a tree continually records everything that happens to it as it grows. Plants are very different from us, but they begin life the same way we do, with a sperm and an egg. People think of trees as objects, just something by the side of the road, like a rock. Trees are responsive and alive. They react more slowly than we do, but see how intricate they become. Kronos started from a seed as big as a finger-nail clipping, and it did *this*."

We were surrounded by buttresses, platforms, and Gothic towers, reaching out of sight. The Atlas Grove is believed to be the oldest grove of redwoods on the planet, and we were aloft in what is thought be its oldest part. Zeus Tree may have been alive when the worship of the god was strong in the cities of Hellas.

"There is a larger issue," Sillett went on. "The redwood forests of California were the most beautiful forests on earth, and they're al-most totally gone. They were reduced to scraps by us. Our society— and I don't mean just American society; I mean Chinese, Brazilian, European society, all of us as humans—we are homogenizing the earth's biosphere. We don't know what will happen to the biosphere or to the forests. I'm afraid that our work trying to understand the redwood forest might just turn out to be documenting something magnificent before it winks out. This forest gives us a glimpse of what the world was like a very long time ago, before humans came into ex-istence. We are in one of the last great rain forests remaining in the temperate zone. These tiny little pockets are all that's left of it. We can talk about conserving biodiversity, conserving species, but that isn't enough. We could keep the redwood species alive as a bunch of little redwood trees, but this forest and all that it shows us would be gone."

"What does it show us?"

"Maybe these trees can teach us something about ourselves. Marie and I and you, we're nothing. We're little snapshots in time, and we'll soon be gone. This grove has burned in huge fires during the past millennia. Redwoods don't die if they burn. A redwood can be burned to a blackened spar, and afterward it goes, 'Wooah,' and just grows back. Look at Kronos. It's been hammered. It's dying. And it's more beautiful than ever. These trees can teach us how we can live. We can be hammered and burned, and we can come back and be more beautiful as we grow." He paused. "Dude, it's getting dark. We need to go down."

WE DROPPED DOWN INTO A MAZE OF STANDiNG TRUNKS OF ALL SIZES. It was an aerial grove of redwoods that rise out of a buttressed platform extending from the side of Kronos, the largest trunk complex growing from a limb that has yet been identified on a redwood. It was the Great Kronos Complex, otherwise known as Kronos Wood. It had twenty-two trunks in it, springing out of a huge mass that grew sideways from the tree's main trunk. The platform extended for sixty feet out of the side of Kronos. The bigger "trees" in Kronos Wood were between eighty and a hundred feet tall and up to a yard and a half across at their bases. The largest of them was bigger than the Red Queen, the biggest tree in my woods.

We descended through Kronos Wood, spidering down along different trunks and pulling tape and tags off as we went. Kronos Wood was a fractal maze of redwoods springing from redwoods. It reminded me of a remark by the physicist Eugene Wigner, who said that the fact that nature seems to know mathematics is a wonderful gift that we neither understand nor deserve.

We ended up standing in a bed of earth and ferns on the floor of Kronos Wood. We tidied up the place, taking out the salamander lodges. Then Sillett stepped into the air and disappeared, rappelling down into the tops of some very tall hemlock trees that were not tall enough to reach the base of Kronos Wood.

"Oh!" his voice rang up.

"What is it, Steve?" Antoine asked.

"My rope isn't long enough. I haven't reached the ground."

He was dangling at the very end of his longest rope, twenty feet above the forest floor.

I leaned out. "What are you going to do?"

No answer. Clinking sounds. A minute later, he said, "I'm on the ground." He had clipped his short spider rope to the longer one, creating a daisy chain. The chain had touched the ground, and he had gone down along it.

I would be the last person to see Rhea Garden and Kronos Wood for a long time, if ever again. Sillett had removed all the climbing aids from the trees. He and Antoine intended never to climb in the west end of the Atlas Grove in their lifetimes, and they intended that no one else would, either. The trees would be left to go wild.

MICHAEL TAYLOR'S DREAM

MICHAEL TAYLOR GREW TIRED OF REPAIRING THE NICO-derm machines and quit that job, and moved, with Conni, to a hamlet in Trinity County, where they bought a few acres of land and a small house. Taylor started trading silver objects on eBay. "I'm buying and selling sterling flatware, tea sets, and silver figurines," he said to me one day in Arcata, when I saw him there.

Taylor had never given up his hope of finding the world's tallest tree. He began experiencing the recurrent dream about finding the Ultimate Tree, the tree that ran clear out of sight. He hadn't had the dream in years, probably not since he had been working at the C&V Market. Now he began dreaming that he was discovering the Ultimate Tree in Redwood National Park. In late 2005, Taylor talked Chris Atkins into exploring certain valleys there more closely. In a series of harrowing bushwhacks they crawled through unexplored places they had never been—pockets that contained virgin fragments of redwood forest outside the Worm, in the tracts that had been added to the national park in 1978.

About halfway into one valley, they discovered a 354-foot-tall redwood, and they named the tree Maia, after the most beautiful of the Pleiades sisters. During the hike out, Taylor became exhausted, and began vomiting and couldn't stop. "Maia is the wrong name," Sillett said to him, later. "You heaved your dinner and your breakfast after you discovered it. I'm naming it the Dry Heaves Tree." The creek where it lived became known as Dry Heaves Creek.

Then Atkins and Taylor announced that they were going to explore the valley I referred to as Fog Canyon. Rarely if ever visited by people, Fog Canyon was very hard to enter on foot. It was a significant place in the emotional life of Taylor and Steve Sillett. This was the place where, years earlier, Steve had broken down and wept, and told his friend that he thought he might die in a fall. Taylor had tried to encourage him then by saying, "Maybe one of the trees we're looking at right now is the world's tallest." Sillett somewhat reluctantly agreed to go on another bushwhack into Fog Canyon, and he, Taylor, and Atkins explored it in the spring of 2006. They turned up nothing in Fog Canyon. Sillett announced that he would never go back to any more of Taylor's hellish places. "Those valleys kick our asses every fricking time we go in there," he complained to Taylor. "To go in there is just self abuse."

JULY 1, 2006, DAWNED CLEAR AND FRESH ON THE NORTH COAST, AND Michael Taylor and Chris Atkins decided to go on another bushwhack into Dry Heaves Creek, despite Sillett's admonitions. The forest was extremely dense, and it took them seven hours to get into the valley. Close to six o'clock, after the sun had gone down behind a ridge, they decided to measure a couple of the trees and then turn back. Then they noticed a top shining in sunlight. All the other trees were in the dark. When they aimed their lasers on the sunlit top, they found that the tree was taller than the Stratosphere Giant. The redwood was about 375 feet tall. It was the world's tallest known tree. They put their arms around each other. "I knew it was here," Taylor said to Atkins. "I dreamed about this tree."

They named it Helios, after the Greek god of the sun, and the grove became known as the Helios Grove. The grove was magnificent, one of the finest in Redwood National Park, and it had never been discovered. Helios Tree was growing on a steep slope, not in flat ground along the creek, where tall trees would be expected to occur. Minutes after they discovered Helios, they found another tree, a darning needle of a redwood with a dead, sun-bleached top that looked like driftwood, which the laser readings indicated might be 371 feet tall—taller than the Stratosphere Giant but shorter than Helios—making it, perhaps, the world's second-tallest tree. They named it Icarus. "Icarus flew too close to the sun and got burned," Taylor explained.

Sillett apologized to Taylor for his remarks about dry heaves, and made plans for the first ascent of Helios. Then, on August 25, while Sillett was still planning the climb, Atkins and Taylor penetrated Fog Canyon beyond the place where Sillett had wept ten years earlier. Not far from the hillside where Taylor had suggested to Sillett that they might be looking at the world's tallest tree, they discovered exactly that. A redwood a lot taller than Helios. It, too, was growing on a slope, in one of the hidden patches of primeval redwoods, missed by the loggers, that had been added to the national park in 1978. They named the tree Hyperion, after the Titan who was the father of Helios. Fog Canyon would be named Hyperion Valley. The valley might not have had human visitors pass through its length in thirty years. (Loggers had probably scouted through the valley before then.) Using their lasers, they estimated that Hyperion was close to 380 feet tall. This was stunning. "I am truly honored to be part of this journey with you and Chris," Sillett told Taylor. "Thank you for bringing such joy to my life." Chris Atkins and Michael Taylor had discovered one of the crown jewels of nature on earth.

STEVE SILLETT, MARIE ANTOINE, JIM SPICKLER, AND I BUSHWHACKED into Hyperion Valley on the first day of climbing season, in September, wearing backpacks full of climbing gear. Chris Atkins and Michael Taylor moved along with us. We were accompanied by three officials from Save-the-Redwoods League, a television crew with *National Geographic,* and the chief ranger of Redwood National Park, a woman named Pat Grediagin. Among park employees, the location of Hyperion was supposed to be known only to Grediagin, for the time being. "There's been a lot of talk about this discovery," Grediagin said. "I'm just worried someone will get a wild idea to try to find this tree."

She knew of a good way into the valley. Even so, we traveled for what seemed like hours, passing stumps that were ten to fifteen feet across and were bitten with chainsaw marks. The stumps were buried in thickets of redwoods no bigger than bean stakes, mixed in with alder trees. We passed through a kind of keyhole in the landscape and entered Hyperion Valley. We began wading upstream along a creek. The only way to get into the valley was to stay in the water. We hunched to get through salmonberry brambles. We could see the sky, since the big redwoods were gone from this part of the valley. The creek ran clear and cold, and its pools were populated with red-legged frogs—*Rana aurora,* a denizen of the rain forest. We came to a pile of redwood trunks, blocking the valley and rising far above our heads. We wriggled through the pile, passing our packs along from hand to hand. "I call this 'groveling' to get into the redwoods," Sillett said.

Beyond the pile, we entered the lost caverns of Hyperion Valley. The sunlight faded, sounds grew soft, and the forest became serenely vertical. We passed into the Hyperion Grove of Redwood National Park, an unknown place in the world until three weeks earlier. Big laurel trees packed the understory, filling the air with a spicy aroma. They seemed like shrubbery under the giants. The branches of the laurels were draped with isothecium moss—a wispy rain-forest moss that hangs down in beards. We climbed through walls of sword ferns, and suddenly came upon Hyperion. The tree was fifteen feet across near the base, a giant column leading into the green undersurface of the canopy.

This was to be the first ascent. Sillett fired a crossbow into the tree, while the *National Geographic* crew filmed him. Michael Taylor sat by himself up on the hillside, his back against a redwood log. There was a swish: an arrow flew off Sillett's crossbow, and Taylor's eyes followed it into Hyperion.

"What are you thinking, Michael?" I asked.

"I guess I'm feeling a sense of completion," he replied. "A long time ago, I set out to do something. Now it looks like I've reached the goal. It's like if I die tomorrow, everything will be all right." He fell silent for a moment. "I think I've found out the secret of making a dream come true."

"What's that?"

"Just don't stop. Don't stop. Don't ever stop. If someone tells you something is impossible, do that thing first. Prove that it *is* possible, and keep going."

Sillett went aloft. Marie Antoine followed. Spickler and I put on our gear and waited. "I'm finding tiny golden-brown ants up here," Antoine said on the radio. "It's only the second time I've ever seen ants in a redwood."

"That may be a new species," Sillett said. He was heading for the top.

Spickler and I went aloft, and spent the afternoon swinging in circles around the tree, sometimes arcing twenty or thirty feet out into the air, while we pulled a measuring tape around the trunk at intervals and moved slowly upward. Hyperion turned out to be 379.1 feet tall. It contained 18,600 cubic feet of wood.

Late in the afternoon, I reached the leader of Hyperion and stopped at a place where I could just about wrap my joined hands around the trunk, close to the top. A wind started to flow down the valley as the mountains began to cool. The crown of Hyperion speared into light and began swaying and hissing. The creek poured far below, invisible in velvety black. Other redwoods drifted up, nearly as tall as Hyperion, but they didn't have names. Each tree had a slightly different color. There were deep greens, gray-yellow greens, brown greens, deep-blue greens, and bluish grays. Because the redwood species is so old, the trees have a large amount of genetic variation, and it comes out in the trees' colors.

I put a descender on the rope and went slowly down through Hyperion until I reached the bottom surface of the canopy, twenty stories above the ground. Then I took a last look around, and cast off. I kicked away from the tree as hard as I could and opened the brake on the descender full wide. The rope began to rush through the descender, and I fell out of the canopy on a fast rappel. Huge columns appeared, the trunks of Hyperion Grove, and I floated weightless down through redwood space.

A NOTE ON TREE CLIMBING

If you are inspired to climb trees, please remember that tree climbing, like all forms of climbing above the ground, is inherently dangerous, and can result in severe injury or death. The tree-climbing techniques I've described in this book require expert skills, advanced training, and highly specialized climbing equipment. It is always a climber's personal responsibility to get proper training and to use appropriate safety equipment and techniques. No one should attempt to climb trees without sufficient training from a certified tree-climbing instructor. Tree-climbing instruction is available through several organizations.

GLOSSARY

AERIAL TRAVERSE Movement horizontally from tree to tree with ropes. Sometimes done by SKYWALKING, sometimes by ZIPLINE.

ALLUVIAL FLAT A usually small floodplain at the bottom of a valley, near a creek or a river where, often, the largest and tallest redwoods occur.

ANCHOR POINT A strong branch, limb, or crotch in a tree over which a loop of climbing rope is placed.

ANGIOSPERM A flowering plant; it bears flowers and fruits containing seeds.

ARBORIST Formerly known as a tree surgeon, a person who prunes trees, takes care of them, and cuts them down when necessary.

ARBORIST CLIMBING TECHNIQUE The specialized method used by ARBORISTS to climb trees.

BELAY ROPE A safety rope attached to a climber and held by a belayer, who stands below the climber.

BRANCHWALKING Walking very lightly along a branch while keeping one's body suspended from a rope anchored overhead or at a diagonal.

BUTTRESS Dense wood formed below a secondary trunk to support its weight in the crown.

CAMBIUM A layer of wood, the living part of the tree, which exists under the bark, outside the heartwood of the tree.

CANOPY See FOREST CANOPY.

CANOPY BONSAI Small trees growing as EPIPHYTES high up in the crowns of redwoods and other forest trees.

CANOPY-TREKKING Traveling horizontally from tree to tree in a forest canopy while pulling all of one's ropes and equipment along, so that nothing is left behind. See also MOTION, GOING INTO.

CARABINER Metal clip, made of aluminum or steel, used by climbers to attach ropes to places.

COAST REDWOOD *Sequoia sempervirens,* a type of conifer occurring along the California coast. World's tallest tree.

CONIFER A type of tree that includes pines, firs, spruces, hemlocks, cedars, and redwoods, which typically produce seeds in cones. Conifers are GYMNOSPERMS.

CRATERING Dying in a fall from a tree.

CROWN The leafy part of a tree, consisting of its radiant array of branches, limbs, and REITERATED trunks.

DEBRIS RING Ring-shaped or crescent-shaped pile of woody debris created when part or most of a tree falls to the ground around its standing trunk.

DETONATION ZONE Area of devastation when a large tree falls.

DIRECT CLIMBING (of a tree) The climbing of a tree with one's body and ropes.

DISCOVERY (of a tree) This doesn't necessarily mean that a tree has never been seen before. It means that nobody has understood the tree's size or measured it.

EPICORMIC (EPI) BRANCH A branch arising from the scar of a broken branch, often as part of a fan-shaped spray growing from the trunk well below the top of the tree. Many trees have epicormic branches. In redwoods, small epicormic branches are not solidly attached to the trunk and can easily break or fall off the tree.

EPIPHYTE A plant that grows on another plant, typically on the branches of a tree, without parasitizing it. The roots of an epiphyte do not touch the ground.

FAIRY RING A ring of redwoods that may have sprouted from the roots of a cut or burned redwood.

FERN MAT Fern garden in a redwood occurring on limbs and in crotches.

FIRE CAVE Cave in a redwood carved by a fire.

FOREST CANOPY Defined by biologists as the part of a forest that is above the level of a person's eyes; thus, all of a forest that is more than about five feet above the ground.

FREE-CLIMBING Climbing without ropes or safety gear.

GIANT SEQUOIA *Sequoiadendron giganteum,* a type of conifer occurring in a few spots in the western Sierra Nevada of California.

GYMNOSPERM A type of plant that bears seeds but not flowers or fruits.

HECTARE Ten thousand square meters of ground, or an area one hundred meters on a side. Equivalent to two and a half acres.

JUG, JUGGING Act of climbing up a vertical rope using a pair of mechanical rope ascenders, handheld devices with teeth that grab a rope.

LEADER Highest thrust of living wood on a tree.

LICHEN An organism composed of a fungus living in association with an alga or a cyanobacterium (photosynthetic bacterium). Lichens typically grow on rocks or on the bark and branches of trees.

MAIN ROPE A long rope going from the ground into the crown of a tree along which climbers move up and down.

MILLENNIAL STRUCTURE A structure on a very old tree that appears to have taken at least a thousand years to form.

MOTION, GOING INTO When a climber detaches from the MAIN ROPE in a tree and moves freely through the CROWN, or performs AERIAL TRAVERSES from tree to tree. See also CANOPY-TREKKING.

MOUNTAIN ASH *Eucalyptus regnans,* the tallest tree in the Southern Hemisphere. Occurs in southeastern Australia (in Victoria and Tasmania).

NINJA CLIMB Climb of a tree without official permission. A ninja tree climber often uses dark clothing and equipment, and may climb at night.

OLD-GROWTH FOREST Forest in which the trees have been growing undisturbed for long periods of time. The exact meaning of the term is much debated among forest ecologists.

PHOTOSYNTHESIS Sunlight-driven metabolic process in plants and other kinds of organisms involving the consumption of carbon dioxide and the release of oxygen. It produces sugars, which the plant can turn into cellulose, starch, and proteins.

RAIN FOREST A wet forest that typically receives at least eighty inches of rain a year. Temperate rain forest occurs in cool but not cold climates, and typically consists of CONIFERS, which are GYMNOSPERMS. Tropical rain forest occurs in hot climates and typically consists of ANGIOSPERM (flowering) trees.

RAPPEL To slide down a rope using a descender device, also called a rappelling device.

REDLINE A height of fifty feet or more above the ground. A fall from above the redline typically results in death.

REITERATION *Noun*—a smaller part of a tree that looks like the tree itself. *Verb*—the process whereby a tree adds smaller structures that look like the tree itself. Term defined by the French botanist Francis Hallé.

SKYWALKING Traveling in midair horizontally or at a diagonal while hanging suspended from a V of ropes attached to branches overhead.

SPIDER ROPE, SPIDER RIG Also called a motion lanyard or a double-ended split-tail lanyard. A rig of spliced rope tied in special knots, used by climbers to move in three dimensions through the crowns of the world's tallest trees.

TREEBOAT Hammock used for camping in tree crowns.

TREK, TREKKING See CANOPY-TREKKING.

TRUNKWALKING Walking straight up a tree trunk with one's feet planted on the trunk while one's body is held horizontal and is suspended from a rope.

VIRGIN FOREST A forest that has never been logged. See also OLD-GROWTH FOREST.

WILD TREE A tree that has never been climbed and explored.

ZIPLINE Horizontal AERIAL TRAVERSE rope running between two trees.

ACKNOWLEDGMENTS

CANOPY SCIENCE AND TREE CLIMBING: I'm deeply grateful for the generosity of Stephen C. Sillett and Marie Antoine in allowing me to share with them the exhilaration and wonder of exploring the world's tallest forests. Their circle of collaborators welcomed me: Robert Van Pelt, George W. Koch, James C. Spickler (who partnered with me in many climbs in California and Australia, including the first climbs of Hyperion and Helios), Anthony R. Ambrose, Cameron Williams, and Giacomo Renzullo. Michael W. Taylor and Chris K. Atkins devoted many days to showing me their discoveries and, at one point, took me on one of the toughest off-trail backpacking trips imaginable, into remote parts of Redwood National Park, searching for undiscovered giant redwoods. Jim Taylor, Jeb Taylor, Cassie Hendry, and Conni Metcalf generously gave their recollections and time, as did Terence and Julianna Sillett and T. Scott Sillett. Many thanks also to: Marwood Harris, Scott Altenhoff, Jack Popowich, Joe Cordaro, R. Steve Foster, Jon Shaffer, Kevin Hillery, Tom Ness, Sophia Sparks and the staff of New Tribe, and Peter and Patty Jenkins of Tree Climbers International. Amanda LeBrun generously shared her perspective, deepening this book in important ways. With the U.S. National Park Service: Pat Grediagin, Rick Nolan, and Jim Wheeler. In Scotland: Alan Watson Featherstone, the founder of Trees for Life; Malcolm Wield and his staff with the Forestry Commission of Scotland; and Howard and Hilary Johnson of Kerrow House B&B for helping with expedition logistics. I received encouragement and help from the staff of Save-the-Redwoods League: Ruskin K. Hartley, Kenya Lewis, Dorinda Nyberg, and Christine A. Ambrose. Nalini M. Nadkarni and

Margaret D. Lowman gave their perspective and enthusiasm, and have been a continuing inspiration to me. Tim Kovar was my principal climbing instructor; he also trained and climbed with my children.

PUBLISHING: Sharon DeLano, this book's editor, shaped it on all levels, from the architectural level to the level of the comma. This is the fourth book of mine that she has edited. Andrew Joslin, the book's artist, poured himself into the maps and drawings. Joshua Hersh fact-checked this book: any errors of fact are mine alone, but where I got something right, Josh was often involved. Daniel Menaker of Random House made key editorial contributions, and Amy Davidson at *The New Yorker* deftly edited the initial pieces, "Climbing the Redwoods" and "Tall for Its Age." Jonathan Karp provided early guidance. I'm most grateful to Lynn Nesbit for agenting this book, along with Tina Bennett, Stefanie Lieberman, Bennett Ashley, and Cullen Stanley. At Random House, I'm grateful to Susan Turner for her beautiful design of this book, to Carole Lowenstein for leading the design of all my Random House books, to Evan Camfield, this book's production editor, and to Jennifer Jones and London King. I'm so honored by all the staff of Random House, past and present, who have so enthusiastically worked on and supported my books over the years.

FAMILY: Michelle P. Preston, my wife, gave me endless encouragement as well as important editorial guidance, and together we soared both intellectually and, on a few occasions, physically into the canopy. Our children, Oliver, Laura, and Marguerite, also climbed. My brothers, Douglas Preston and David Preston, M.D., and their families kindly submitted to the torture of climbing trees with me. Even our parents, Doffy and Jerry Preston, went up into trees on ropes with their grandchildren. I wish to thank my parents-in-law, John and Diana Parham, for their encouragement, and warn them that if they aren't careful they may end up in a tree with me and their grandchildren one day, too.

RICHARD PRESTON is the bestselling author of *The Hot Zone, The Demon in the Freezer,* and the novel *The Cobra Event.* A writer for *The New Yorker* since 1985, Preston is the only nondoctor to have received the Centers for Disease Control's Champion of Prevention Award. He also holds an award from the American Institute of Physics. He lives in New Jersey with his wife and three children.

www.richardpreston.net

Read on for a preview of
Richard Preston's new book,

Panic in Level 4

ONE DAY I WAS INTERVIEWING THE COMMANDER OF FORT Detrick. We were getting toward the end of the interview. I decided to ask one final question. "I'd like to try to convey to readers what it really feels like to be face-to-face with Ebola virus." I said. "So I'm wondering, could I go into Level 4?"

"That should be no problem," the commander answered. As a matter of fact, I was in luck: Hot Suite AA-4—one of the Ebola labs—was "down and cold," he said. "We'll get you outfitted in a blue suit and we'll give you a walk-through."

"What do you mean when you say the lab is 'down and cold'?" I asked.

He explained that the rooms had been completely sterilized with gas and then opened up for routine maintenance. So the rooms weren't dangerous at the moment. Anyone could go into that lab without wearing a space suit; in fact, there were regular maintenance workers in there right now. The hot freezers, too, had been moved out of the lab. Therefore, the lab was completely cold. It was, in effect, a mock-up of a hot zone.

"That's not really what I had in mind," I said.

"What did you have in mind?" he asked.

"I would like to experience the real thing, so that I can describe it better. I'd like to go into a hot BL-4 lab and see how the scientists work with real Ebola."

"That's just not possible," he answered immediately.

Security at USAMRIID was extremely tight. Even so, it was not as tight as it would become. That day in the commander's office was some nine years before the anthrax terror attacks of the autumn of 2001, shortly after 9/11. The anthrax attacks came to be known as the Amerithrax terror event, after the FBI's name for the case. Small quantities of pure, powdered spores of anthrax—a natural bacterium that has been developed into a very powerful bioweapon—were placed in envelopes and mailed to several media organizations and to the offices of two United States senators. Five people died after inhaling the spores, while others became critically ill; some of those people have never fully recovered. Their health and lives were severely damaged. For the most part the victims, including African Americans and recent immigrants to the United States, were low-level employees of the U.S. Post Office who were just doing their jobs. No one has been charged with the Amerithrax crimes. The evidence suggests they were done by a serial killer or killers who fully intended to murder people and may have taken pleasure in causing the deaths. The case remains open.

Officials at the United States Department of Justice named Stephen J. Hatfill, a former researcher at USAMRIID, as a "person of interest" in the case. Hatfill has never been charged with involvement in the crimes, though. At the same time, there was speculation in the news media that the exact strain of anthrax used in the attacks might have come out of an Army lab, even possibly from USAMRIID itself, where defensive medical research in anthrax had been going on for years. (The precise results of the FBI's analysis of the anthrax strain have not been disclosed by investigators, as of this writing.) USAMRIID scientists, in fact, played a key role in the forensic analysis of the anthrax that was collected from the envelopes.

Following the Amerithrax terror event, security at USAMRIID became astronomically tight. After that, it would have been useless for a journalist to ask to go into the labs. Back at the time when I was researching *The Hot Zone,* though, there was a slight amount of flexibility in the policy. On certain occasions, the Army *had* allowed untrained or inexperienced visitors to go into hot zones at USAMRIID. Unfortunately, as the commander explained to me, some of these visits had ended badly. He said that people who were not familiar with space-suit work in a biohazardous environment had a tendency to panic in Level 4.

In one such an incident, a medical doctor—a visitor—who had apparently never worn a biohazard space suit attended a human autopsy in a Level 4 morgue at the Institute. This hot morgue is called the Submarine. The Submarine is a sealed hot zone with an autopsy room and an autopsy table. The cadaver was believed to be infected with a Level 4 Unknown X virus. During the examination, while the space-suited autopsy team was removing organs from the cadaver, some members of the team noticed that the visiting doctor's face seemed red. As the team members looked at the visitor through his faceplate, they saw that his face was also dripping with sweat. Meanwhile, the outer surfaces of his space-suit gloves and sleeves were smeared with blood from the cadaver.

Reportedly, the man began saying, "Get me out!" Suddenly he tore off his helmet and ripped open his space suit, gasping for breath, taking in lungfuls of air from the hot morgue.

The members of the autopsy team tried to calm him, and they took hold of him and hurried him to an airlock door leading to the exit. They opened the door, pushed him into the airlock. At least one of the team members accompanied him into the airlock. The airlock was closed, and the chemical shower was started.

The way I heard the story, the man stood or sat in the airlock while the chemicals ran over his head and neck and down inside his opened space suit. The shower stopped automatically after seven minutes. The chemicals had flooded his suit. Then the team members helped him into the staging area—the so-called Level 3 area—and helped pull him out of his space suit. By this time, he was subdued and embarrassed.

At USAMRIID, people who have had a verified exposure to a hot agent are put into a Level 4 quarantine hospital suite called the Slammer. The Slammer is a biocontainment unit where doctors and nurses wearing space suits can treat a patient without being exposed to a virus the patient may have. The man who had panicked was a possible candidate for quarantine in the Slammer. Even so, after an investigation, the Army felt that he did not need to be put in quarantine; there was no evidence that the cadaver had actually been infected with a virus. And the man never got sick.

"We can't predict how someone who's untrained might react in BL-4, so we can't allow you to go in," the commander explained to me.

I still wanted to go into Level 4. But I couldn't see how to get there.